THE GULF STREAM

A line of sargassum weed running parallel to the inshore edge of the Gulf Stream at 40° N., 63° W. There is usually an abrupt change of velocity of half a knot or so on both sides of a weed line such as this. Often there are a number of such lines, and corresponding velocity steps, along the inshore edge of the Stream. The vessel is about 300 feet long. This aerial photograph was given to me by Commander William Kielhorn, United States Coast Guard Reserve.

The
GULF STREAM

A Physical and Dynamical Description

SECOND EDITION

BY HENRY STOMMEL

University of California Press

Berkeley, Los Angeles, London

UNIVERSITY OF CALIFORNIA PRESS
Berkeley and Los Angeles, California

UNIVERSITY OF CALIFORNIA PRESS, LTD.
London, England

SECOND EDITION
Third Printing, 1972

ISBN: 0-520-01223-2
Library of Congress Catalog Card Number: 64-23710

PREFACE

The purpose of this book is to describe and explain what is known about the Gulf Stream in a way which will interest physical scientists. The name 'Gulf Stream' is familiar enough to everyone, but few scientists have any knowledge of the nature of this grand natural phenomenon. I hope that by means of this book I shall be able to communicate the facts and theories concerning the Gulf Stream to a wide scientific audience. Those interested in an authoritative treatise which covers the entire field of oceanography will do best to refer to *The Oceans: Their Physics, Chemistry and General Biology*, by H. U. Sverdrup, Martin Johnson, and Richard Fleming; those who seek an appreciation of the theoretical framework of oceanography should read *Dynamical Oceanography*, by Joseph Proudman. So far as the restricted subject of the Gulf Stream itself is concerned, I believe the discussion in this work is more comprehensive than that in any other source. At any rate, I have sought to make it so.

I am very much indebted to all my colleagues at the Woods Hole Oceanographic Institution, many of whom have spent weary months at sea in our uncomfortable little exploring vessels gathering information about the western North Atlantic. In particular, I have profited from discussion of a descriptive nature with Messrs F. C. Fuglister, C. O'D. Iselin, W. S. von Arx, A. H. Woodcock, and L. V. Worthington. My collaboration with Mr Donald Parson, Jr., will always evoke memories of happy instrument-making. I have had the pleasure of theoretical discussions with Dr Willem Malkus, Dr Jule G. Charney, Dr George Veronis, and Dr George W. Morgan. Some of the ideas expressed in this book arose and took shape in these informal discussions. I do not claim them as my own. Wherever recollection permits I have indicated their source. And I want to express my gratitude to Dr R. S.

Arthur, of the Scripps Institution of Oceanography, for his advice and encouragement.

In a field as small as dynamical oceanography it is inevitable that many exchanges of information take place by word of mouth. It is always a pleasure to discuss problems freely with investigators from other countries, and I have been able to talk with most of them. Owing to world conditions, I have never met the leading Russian dynamical oceanographer, Professor W. B. Stockmann, and hence my acquaintance with his ideas is limited to occasional translations of his papers. The Japanese studies of the Kuroshio are, for the most part, not treated in this book, partly on account of language difficulties, and partly because they are voluminous and deserve more careful study than I can give them. The Kuroshio is not unlike the Gulf Stream in many ways, and it is hoped that some day a thorough comparison can be made.

Over the course of the years my researches have been generously supported by the Office of Naval Research through contracts with the Woods Hole Oceanographic Institution. Without this sustenance I could certainly not have undertaken such experimental and observational studies as I have, and I should not have been afforded the opportunity of associating with the many keen minds and eager students of the sea who work at Woods Hole. But the conception of and responsibility for a book of this kind is, in the last analysis, an individual matter; therefore, it is also befitting to record here the fact that the writing of this book has been entirely a private undertaking at home and that preparation of the manuscript and of the original figures has been at my own expense.

HENRY STOMMEL

Woods Hole, Massachusetts
June, 1955

CONTENTS

Chapter One

HISTORICAL INTRODUCTION

From the time of the recorded discovery of the Gulf Stream to the present there have been many ideas about its cause. The historical references[1] I shall make will be restricted to those for which some documentary evidence exists—a plan that excludes a large body of speculation about early Norse, Arabian, and Portuguese navigators.

EARLY IDEAS AND EXPLORATIONS

Although the island of Cuba was first circumnavigated in 1508, it was not until 1513 that the Gulf Stream (specifically, the Florida Current) was described by Ponce de León, who, sailing from Puerto Rico, crossed the stream north of Cape Canaveral and then sailed south to Tortugas. The current was so swift that his three ships were frequently unable to stem it (see Herrera y Tordesillas, 1601).

By 1515 Peter Martyr of Anghiera reported various conjectures about the Gulf Stream (see the 1577 translation of his *Decades*). His arguments were based essentially upon the principle of the conservation of mass and upon the tacit assumption that the current velocity is independent of depth. Peter Martyr argued that the North Equatorial Current must either (i) pile up large masses of water at the Brazilian coast, or (ii) pass through some great straits or passages into the Pacific and thence round again into the Atlantic, or (iii) be deflected by the mainland so as to flow back into the ocean (Kohl, 1868). The first possibility was ruled out, he

[1] The material in this chapter is drawn from an article by the author in the *Scientific Monthly* for April, 1950.

claimed, because the explorers of the Brazilian coast had never noticed such a piling up of water. The second possibility was doubtful, because it was the consensus among the Spanish navigators that the mainland was not open, but presented a continuous barrier to the westward flow. Hence, by elimination, only the third possibility was left; and as an example of a deflection of the Equatorial Current by the mainland, the Gulf Stream, said Peter Martyr, was a case in point.

The westward flow of the Equatorial Current itself was usually attributed to the *primum mobile*—in some manner not clearly understood the general westward motion of the celestial bodies across the sky drew the water and air of the equatorial regions along with it.

By 1519 the Gulf Stream was so well known that Spanish ships bound for America came by way of the Equatorial Current but, on their return, passed through the Florida Straits, followed the Gulf Stream to about the latitude of Cape Hatteras, and then sailed eastward to Spain. In this way they had favorable winds and avoided contrary currents over the whole voyage.

The sixteenth century marked the beginning of a period of intense activity in the western North Atlantic Ocean, of exploration of the coasts, and of the search for the Northwest Passage. Many of the early cruises in and about the Gulf Stream, chronicled in Kohl (1868), need not concern us here. Navigators of various nations investigated the geographic extent of the Gulf Stream System. Among these we may number Sir Humphrey Gilbert, who first suggested using a deep-sea anchor to determine surface drift, Martin Frobisher, Ribault, and Laudonnière. Frobisher and John Davis made numerous observations of the Labrador Current.

Toward the end of the century André Thevet (1575) attributed the Gulf Stream to the great rivers that flow into the Gulf of Mexico. It was not until many years later, when the transports of mass through the mouth of the Mississippi and the Straits of Florida were measured and compared, that the utter inadequacy of such an explanation was completely revealed. The mass flux through the former is only one one-thousandth of that through the latter.

The sharp line of demarcation between the warm- and cold-water masses was apparently first recorded by Lescarbot in 1609. His comment, as quoted (from 2d ed., 1612, 2:531) by Kohl (1868, p. 68), reads:

> I have found something remarkable upon which a natural
> philosopher should meditate. On the 18th of June, 1606, in
> latitude 45° at a distance of six times twenty leagues east of the
> Newfoundland Banks, we found ourselves in the midst of very warm
> water despite the fact that the air was cold. But on the 21st of

June all of a sudden we were in so cold a fog that it seemed like
January and the sea was extremely cold too. [Translation.]

In 1590 John White took a trip from Florida to Virginia. In order to
stay within the Gulf Stream, he reported, one had to stand far out to sea,
because along the coast there was a countercurrent—'eddy currents setting
to the south and southwest' (Kohl, 1868). This was the first mention of
countercurrents on the shoreward side of the Gulf Stream.

The seventeenth century saw the colonization of the Atlantic coast of
North America, and the Gulf Stream was of course traversed countless
times at various latitudes. A number of studies of ocean currents were
published at this time, works of most varied quality. Varenius (in 1671;
2d ed., 1681) published a very comprehensive description of the surface
currents then known, and Isaac Vossius (1663) postulated a complete
circulation of the North Atlantic Ocean, turning in an ocean-wide clockwise motion. The first chart showing the Gulf Stream was Kircher's,
published in 1665 (3d ed., 1678); and the next was a current chart by
Happelius, in 1685 (see Kohl, 1868). In addition to showing certain true
features, these charts display certain extraordinary phenomena—for
example, two distinct surface currents which cross over each other; and
a great whirlpool off the Lofoten Islands, the legendary Maelstrom. The
reader can find some of these charts reproduced in Pillsbury's (1891)
account of his studies. On the whole, these charts were far superior to the
theories advanced to explain them. Vossius contended that a great mountain of water was formed each day at the equator by the heat of the sun,
and that this water mass was carried westward and broke upon the
American shore, and then flowed along the coasts in the form of currents.
Kircher, it is true, suggested that the trade winds contributed to the ocean
circulation, but he also enumerated other, fantastic, causes. Even well-
informed men like Kepler had curious ideas about the causes of ocean
currents. Kepler believed that because the water is only loosely attached
to the earth it could not keep up with the diurnal rotation and hence fell
behind, the result being the westward drift of the Equatorial Current
(Kohl, 1868, p. 87).

These various theories were, at least, honest attempts to explain physical
phenomena by an as yet poorly developed physics. In addition, there were
advanced fantastic theories that enjoyed a certain popularity. An example
is Merula's statement (Kohl, 1868, p. 63):

At the North Pole one finds four large islands...between which
are four deep and broad channels. The water flows together near
the Pole, but at the Pole itself is a great Black Rock, 33 leagues

in circumference. Ships which once enter one of these channels never return, not even with the most favourable winds, and next to the Black Rock all the water is engulfed into the bowels of the earth, whence it flows through springs and river sources once again into the light of day. [Translation.]

From the *Louisiane* in 1702 Laval observed what he thought were variations in the strength of the Florida Current associated with the north component of wind; and in his book about the voyage published some years later (1728), he stated that he believed this phenomenon to be common knowledge among sailors.

THE PERIOD 1700–1850

During the early 1700's the great American whale fishery sent ships all over the world, and the names of such patches of sand as Nantucket became known in every land touching the sea. The practical knowledge gained by these seafarers was not widely published in technical journals; instead, it was handed down by a system of apprenticeship and by word of mouth. And meanwhile, the basic understanding of fluid mechanics was growing. Daniel Bernoulli's *Hydrodynamica* was published in 1738. In the last third of the century the study of the theoretical aspects of oceanic tides was brought to a high point in the contributions of Laplace. The influence of these developments in theoretical mechanics, and the general intellectual atmosphere of this Age of Enlightenment, discouraged further extraphysical and purely imaginative theories of ocean currents.

In 1770 the Board of Customs at Boston complained to the Lords of the Treasury at London that the mail packets usually required two weeks longer to make the trip from England to New England than did the merchant ships. Benjamin Franklin was Postmaster General at the time and happened to discuss the matter with a Nantucket sea captain, Timothy Folger. The captain said he believed that charge to be true (Franklin, 1786, p. 314):

> 'We are well acquainted with the stream because in our pursuit of whales, which keep to the sides of it but are not met within it, we run along the side and frequently cross it to change our side, and in crossing it have sometimes met and spoke with those packets who were in the middle of it and stemming it. We have informed them that they were stemming a current that was against them to the value of three miles an hour and advised them to cross it, but they were too wise to be councelled [*sic*] by simple American fishermen.'

Franklin had Folger plot the course of the Gulf Stream for him and then had a chart engraved and printed by the General Post Office.

Franklin (1786, p. 315) believed that the Gulf Stream was caused

> by the accumulation of water on the eastern coast of America between the tropics by the trade winds. It is known that a large piece of water 10 miles broad and generally only 3 feet deep, has, by a strong wind, had its water driven to one side and sustained so as to become 6 feet deep while the windward side was laid dry. This may give some idea of the quantity heaped up by the American coast, and the reason of its running down in a strong current through the islands into the Gulf of Mexico and from thence proceeding along the coasts and banks of Newfoundland where it turns off towards and runs down through the Western Islands.

By the time of Maury, in the middle of the nineteenth century (see Maury, 1859), Franklin's estimates of the velocities of the Stream were regarded as excessive, but more recent studies tend to confirm them. Franklin did not give any details concerning the edge of the Stream. Starting in 1775, both Franklin and Charles Blagden (1782), independently, conceived the idea of using the thermometer as an instrument of navigation, and each made a series of surface temperature measurements while crossing the Atlantic. On Franklin's last voyage in 1785 he even attempted to measure subsurface temperatures to a depth of about 100 ft., first with a bottle and later with a cask fitted with a valve at each end.

A number of subsequent investigators made use of the surface thermometer, among them Governor Pownall (1787) and Captain Strickland (1802). It was in this fashion that Captain Strickland discovered a north-easterly extension of the Gulf Stream toward England and Scandinavia. These temperature measurements were not made with any idea of determining the pressure field and geostrophic current, as is done today, but were simply regarded as an indication of the type of water through which the ship was sailing.

In passing, the invention of the marine chronometer by John Harrison and its perfection by Thomas Earnshaw should be mentioned. By 1785, accurate chronometers were generally available to ships; this made the determination of longitude at sea at last a possibility and the determination of the set of a current much more exact. Another important oceanographic tool, the drift bottle, was probably first used in 1802, when such a bottle was cast from the *Rainbow*. The use of drift bottles continued for more than a century, and finally received great impetus at the hands of the Prince of Monaco.

One important consequence of these early thermometric measurements was the discovery of pockets of cold water in the Gulf Stream. These pockets were first observed on the noteworthy cruise of the packet *Eliza*, en route from Halifax to England in April, 1810 (see Kohl, 1868). In the middle of the warm water of the Gulf Stream a mass of water colder than the surrounding water by about 10–15° Fahrenheit, and some 200 miles in diameter, was discovered. The explanation offered at the time was that these cold spots were due to the melting of icebergs entrapped by the Gulf Stream. This occurrence suggests a phenomenon very much like the large cold eddy reported in recent times by Iselin and Fuglister (1948), but its real nature and cause must remain in the realm of surmise. Another view, much in accord with modern synoptic experience, was that held by John Hamilton, who made serial air and water temperature measurements during twenty-six voyages to and from Europe, and asserted that the Gulf Stream was so unsteady and shifted its position so frequently that it was impossible to define its limits. Both Humboldt (1814) and Sabine (1825) were convinced that changes in the strength of the trades affected the Gulf Stream, and Sabine even suggested the use of weather ships which, he said, should observe the Florida Current and then sail quickly to Europe with the news of how strongly it was flowing, so that weather predictions could be made.

In 1832 the results of an extensive compilation by James Rennell of data from the British Admiralty Office were published posthumously. Rennell distinguished clearly between 'drift currents', which are produced by the direct stress of the wind, and what he called 'stream currents', which are produced by a horizontal pressure gradient in the direction of flow. Rennell, in accord with Franklin's earlier view, regarded the Gulf Stream as a current of the second kind and decided from his investigation that: (i) the breadth of the Stream changes from time to time; (ii) the breadth can vary as much as twofold even within so short a period as ten weeks; (iii) the variations are not seasonal; (iv) the north side of the Stream is more permanent than the south side; (v) temperature alone does not prove the existence of the Stream, for even warm countercurrents may exist; and (vi) cold-water inclusions occur within the body of the warm water. Also, Rennell proposed a special nomenclature for various parts of the Gulf Stream System.

The work of Rennell was so authoritative and exhaustive at the time that it must have seemed to his contemporaries that the major features of the Gulf Stream had been delineated and charted. Few seriously entertained doubts that the Gulf Stream was a result of the downhill flow of water piled up by the trade winds along the American coast and in the Gulf of Mexico.

At this point a discovery was made which for a time completely discredited the notion that the winds are the cause of the ocean surface currents. Arago drew attention in 1836 to the results of a leveling survey across Florida which showed a difference of not more than 7½ in., and probably less. This difference of level seemed much too small to produce the Gulf Stream; hence Arago advanced the idea that the cause of the ocean currents is simply the density differences at the equator and poles due to unequal solar heating. 'We should use the same theory for ocean currents which we use to explain the Trade Winds', he stated. Seafarers were still inclined to the theory of wind-driven currents. Our present ideas of the magnitude of frictional forces in the Gulf Stream suggest, however, that a head of 7½ in. is adequate to drive the Stream.

THE LAST HALF OF THE NINETEENTH CENTURY

In 1844 an intensive study of ocean circulation was begun by Matthew Fontaine Maury (1859), who was partial to Arago's theory. He dismissed Rennell's theory of wind-driven circulation on the following grounds: (i) assuming that the transport through the Florida Straits is the same as that of the current off Cape Hatteras, he (Maury) deduced the depth at Hatteras, completely ignoring the possibility that the Stream may mix with its environment; (ii) since the depth deduced in this manner is less than that at the Straits of Florida, Maury made the naïve claim that the Gulf Stream would have to run uphill, and this reduced the wind theory to absurdity. He did not realize that the horizontal pressure gradient along the axis of the Stream is not controlled by the slope of the bottom boundary, so that if the top surface slopes down toward the north, the current can very well flow northward despite its diminishing depth. Maury was very much confused concerning fluid mechanics, even though that science was being rapidly advanced in his time; Stokes's famous paper on the flow of viscous fluids was published in 1845, and the significant work by Coriolis had appeared in 1835. Maury uses such examples as the accumulation of sawdust in the center of a basin filled with water to explain the accumulation of *sargassum* in the Sargasso Sea, and, to illustrate the Coriolis force, the fact (?) that railroad trains usually run off the rails to the right— examples having no more than extremely doubtful applicability to the ocean. He goes on to ask why, if the Atlantic circulation is a completely closed circuit, there should be a piling up at any one place. This shows that Maury did not understand Rennell's distinction between drift currents and stream currents. There seems to be little point in enumerating Maury's misconceptions about the effect of the earth's rotation on ocean currents, his roof-shaped current theory, his strange idea that waters of different salinities are immiscible, and his suggestion that there is a kind of peristaltic driving force acting upon the Gulf Stream. Maury says that the difference of density at the poles and the equator must produce the ocean currents, but he had no quantitative idea about how salinity and temperature are related to density. Accurate hydrographic tables had not yet been constructed, and the salinity distribution in the sea was only imperfectly known.

The encouragement which Maury gave to the collection of ships' data, however, his calling of the Maritime Conference at Brussels in 1853, and his dissemination of hydrographic information, in the form of good pilot charts and revised sailing directions, were of great practical and commercial importance.

Modern surveying of the Gulf Stream began in 1844 with the work done by the United States Coast and Geodetic Survey under the direction of Franklin's great-grandson, A. D. Bache (1860): fourteen temperature sections between Tortugas and Nantucket. These early survey cruises confirmed the existence of cold veins within the Stream. Bache supposed that the cold veins were a result of the bottom configuration which diverted the Gulf Stream in separate bands—irregularities which are now regarded not as straight bands, but as meanders. In Bache's time the 'bands' of cold water were supposed to be invariable in number and position, and the USCGS Gulf Stream Chart of 1860 shows them so. The currents were observed only at the surface by the drift of the ship.

During the Civil War there was a lull in Gulf Stream investigations, but by 1867 they were resumed by Henry Mitchell, who attempted to measure subsurface currents by means of two floats attached by a fine wire, one of which remained at the surface while the other sank slowly. Mitchell concluded that the velocity of the Gulf Stream off Fort Chorrera, Cuba, extended undiminished to a depth of at least 600 fathoms. For the next ten years the *Bibb*, the *Bache*, and the *Blake* continued to survey the Stream. Early soundings had been made by rope, but in 1881 Bartlett made use of the Thomson sounding machine, in which piano wire was used. Bartlett did not detect the cold and warm bands shown on Bache's chart. According to the modern view of the temporary nature of cold inclusions in the Gulf Stream, this is not surprising. The *Challenger*, on her world-wide cruise, visited the Gulf Stream in 1873.

John Elliott Pillsbury (1891) commanded the *Blake* during a remarkable series of observations beginning in 1885. The ship was anchored along several cross sections in the Florida Straits, and the current vector and the temperature at various depths were determined. The anchoring gear and the current meter were of Pillsbury's own design. Pillsbury went about his task in a careful and painstaking manner. For example, he took two years to occupy the section from Fowey Rocks to Gun Cay, 'the time actually employed in observations at Section A being over 1,100 hours'. Observations were also made at four other sections across the Florida Straits, at a station off Cape Hatteras, and at a number of stations in the passages of the Windward Islands.

Pillsbury noted the westward intensification of streamlines within the Straits themselves, as earlier reported by Fremont. By sounding across the Gulf Stream all the way from the Straits to Cape Hatteras and showing that the bottom in this region was comparatively smooth, he disproved Bache's hypothesis that mountain ranges under the sea cause 'bands' of cold water. His observations of temperature and current in the Florida Straits are unique and are so valuable that they are still used (Wüst, 1924) as the classical example of the accuracy of the geostrophic current determination. He was a firm believer in the theory of wind-driven ocean currents. He felt that the time variations of temperature, horizontally and vertically, along the Gulf Stream were abundantly proved, and he attributed them to random local winds and tidal influences along the coast. He established that the Antilles Current is a tributary to the Gulf Stream. He emphasized the importance of the prevailing westerlies in maintaining the North Atlantic Drift. He surmised that the Stream becomes increasingly meandering as it flows northeastward.

During the 1870's there was a good deal of discussion on a qualitative plane between adherents and antagonists of the wind theory. The reader can form a fair idea of the level at which these discussions were carried on by reading the views of Aitken (1877), Carpenter (1874), and Sir C. Wyville Thomson (1874). A mathematical physicist, Zöppritz (1878), attempted to show that the wind stress could cause appreciable ocean currents only after hundreds of thousands of years of constantly acting upon the water, because the molecular viscosity of water is so small. The importance of the role of turbulence in the ocean as an agency in the transfer of momentum, as well as of other properties, was not at that time appreciated. In 1883, five years after the work by Zöppritz, Reynolds' famous paper on his experiments on turbulent flow in pipes appeared. The realization of the fact that the ocean is essentially a turbulent regime showed the error in Zöppritz' reasoning.

It is important to realize that thus far in the nineteenth century the influence of the earth's rotation upon ocean currents was only imperfectly understood by oceanographers, although the hydrodynamical equations of a perfect barotropic fluid relative to a rotating sphere had been written down in the previous century by Laplace (1778). The extremely important fact that the Coriolis force is almost everywhere balanced by the horizontal pressure gradients associated with the distribution of mass was not realized, despite the fact that these terms are several orders of magnitude greater than the terms of the driving force and dissipative force. William Ferrel (1882) was one of the first to understand the role of the Coriolis force in the distribution of ocean currents caused by the wind. He derived the relation between the barometric gradient and the velocity of the wind (which is the counterpart of the geostrophic current relationship of oceanography). For some reason these remarkable achievements of Ferrel were overlooked by oceanographers, and therefore his work did not have a direct influence on the development of physical oceanography. The formula for computing ocean currents from the slope of isobaric surfaces was derived by Henrik Mohn (1885).

The first careful study of the possibility of obtaining an approximation to the velocity field in the ocean from an exact knowledge of the mass field was made by Sandström and Helland-Hansen (1903), on the basis of Vilhelm Bjerknes' (1898) circulation theorem. It became evident that the Coriolis force acting upon the Gulf Stream is counterbalanced by a horizontal pressure gradient associated with the mass of lighter water in the Sargasso Sea, and that the Stream is not so much a warm current as a boundary phenomenon associated with, but not caused by, the sudden downward slope of the isotherms toward the center of the ocean. It was recognized that the differences in density across the Stream have nothing to do with the driving of the Stream, but are simply part of an equilibrium brought about indirectly by the stress of the wind. The great step forward represented by this method of dynamic computation—as it came to be called—was that at last oceanographers had begun to use the hydrodynamical equations of motion in their study of the sea, even though only in a piecemeal way.

By the turn of the century the dependence of density of sea water on its temperature, salinity, and pressure had been determined. The ocean-wide distribution of salinity had been charted and plotted, mostly as a result of the pioneering work of Forchhammer (1865). This development came surprisingly late in synoptic oceanography, especially when one considers how much earlier the surface temperatures in most parts of the North Atlantic had been known.

THE TWENTIETH CENTURY

We now enter the modern period of physical oceanography in which the hydrodynamical equations of mean motion are used, although never in their complete form. There are several reasons for using only a few of the terms. First, the equations are nonlinear in their complete form, and therefore the mathematical methods for solving them are unavailable.

Secondly, the parameters depending upon turbulence are very poorly known. Thirdly, even today we do not have nearly as much accurate synoptic information about the Gulf Stream as we desire.

V. W. Ekman (1905) developed a wind-drift theory, taking into account the effect of the earth's rotation, which for the first time made possible computations of the effect of wind stress in transporting water. In Ekman's treatment the ocean is regarded as homogeneous and of infinite extent horizontally. A constant wind blows over this ocean. In deep water the surface-current vector is 45° *cum sole* to the wind and changes direction and diminishes with depth according to the Ekman spiral.

Later, Ekman (1923, 1932, 1939) carried out further investigations of ocean currents in which he discovered certain deflecting effects of the inequalities in the depth of a homogeneous ocean. A qualitative study of the effect of bottom topography upon currents in a heterogeneous ocean has been advanced by Sverdrup, Johnson, and Fleming (1942). Certain of these deflections have been apparently observed (Neumann, 1940) in the northern parts of the Gulf Stream System associated with the Altair cone.

The first intensive surveys of the Gulf Stream System by the research ketch *Atlantis* began in the early 1930's, soon after the establishment of the Woods Hole Oceanographic Institution. This program was carried out under the leadership of C. O'D. Iselin.

From this point on, I shall dispense with the historical narrative in favor of a topical exploration of the present knowledge concerning the Gulf Stream. It is not surprising that even now, after many years of effort, our conception of the Gulf Stream is incomplete. I shall try to show in what respects this knowledge is deficient, as well as those features which we do understand, and also endeavor to explain some of the greatest difficulties encountered in planning and executing surveys.

Chapter Two

METHODS OF OBSERVATION

The obstacles encountered by the collector of hydrographic data pertaining to the Gulf Stream are numerous and formidable. Although many difficulties remain, modern technology has been of great assistance in oceanographic research.

INSTRUMENTS USED IN GULF STREAM WORK

The instruments utilized are not, for the most part, produced exclusively for the study of the Gulf Stream. The reversing bottle, the thermometer, and the bathythermograph are employed in every ocean area. However, the use of these instruments in the Gulf Stream does require special techniques. For example, anchor stations are out of the question in the strongest part of the Stream, and hydrographic stations are usually made by steaming against the surface current to reduce the wire angle in the low-speed water below. Because most readers are not likely to be familiar with these instruments, this chapter offers a brief description of them and of the techniques for their use at sea.

Reversing bottle.—The reversing bottle is made of metal and is fitted with a valve at each end. When lowered to the desired depth it is tripped by a messenger (a weight which slides along the wire). As it turns over, the valves close and a little more than 1 liter of water is sampled. Normally a number of bottles are lowered in series on the same cable.

Salinity is determined by chemical titration, usually on shore. The standard of accuracy is about $\pm 0.02\%$ ($\%$ means parts per thousand).

Reversing thermometers.—The reversing thermometer is mounted on the sampling bottle. The thermometer is so contrived that when it is turned upside down the mercury column breaks. When the thermometer is brought to the surface the reading must be corrected, because the temperature at which the mercury is broken is usually different from that at which it is read on deck.

These thermometers are of two distinct types, which are used in pairs; that is, two thermometers, one of each type, are mounted on the same bottle. One type is protected from the hydrostatic pressure by an exterior glass shell (thermal contact with the outside is maintained by a pool of mercury); the other type is unprotected from the pressure, and hence registers depth because the mercury bulb is squeezed. The deviation amounts to about $0°01$ C. per meter. The standard of accuracy of each type is $\pm 0°01$ C. Reversing thermometers and reversing bottles are designed for use with the ship hove to or steaming very slowly against the current or into the wind. They are necessary for making geostrophic current calculations.

Bathythermograph.—The bathythermograph is an instrument designed for use from a ship under way. A smoked-glass slide is mounted on a spring-loaded bellows so that it moves under variations in hydrostatic pressure. A bourdon thermal element moves a needle over the slide at a right angle to the direction of the pressure-induced motion. The scratch on the smoked slide is therefore a plot of temperature versus depth. The standard of accuracy in a model which works to a depth of 900 ft. is 10 ft., and $\pm 0°2$ C. Models for other depth ranges are also available.

Bathypitotmeter.—A promising device for towing at low speeds, to make simultaneous recordings of the velocity of water relative to the instrument, the temperature, and depth, was devised by Malkus (1953). The record is made on a moving waxed-paper strip. This instrument is capable of making a determination of a complete vertical sounding in about two hours. The long length of cable, which is payed out when the ship is under way, apparently isolates the instrument from the irregular pitch and roll motions of the ship, which normally cause spurious readings on velocity meters suspended vertically from a drifting ship.

Propeller-type current meters.—There are various kinds of current meters of the propeller type, from the early Ekman meter to the recent Watson meter. They record velocity and direction, but they take so long a time to operate that they are not often used for Gulf Stream studies.

All velocity meters measure water velocity relative to the ship, but the ship's velocity relative to the earth is seldom known adequately from navigation, even in areas with good Loran coverage (where radio fixes good to 1/2 mile are available at any time). Therefore there is always an ambiguity concerning the zero point of all velocity measurements made at sea. Even anchored ships or buoys move about at the end of their cable so much as to be unreliable reference points in deep water.

Towed electrodes.—The component of a ship's velocity relative to the earth and perpendicular to the ship's heading may be determined by measuring the electromotive force induced by the earth's magnetic field in a 100 m. length of wire towed astern. The complete vector velocity may be obtained by occasionally altering the ship's course. As developed by William von Arx, this method has proved very valuable for practical survey purposes in indicating the presence and direction of surface currents (see Longuet-Higgins *et al.*, 1954; and von Arx, 1950). The time required is about 5 min. per measurement, because the ship must be maneuvered in directions other than the regular heading.

The chief difficulty of this method is that it does not give uniformly accurate surface velocities. The reason for this deficiency can be easily made clear by a concrete example. Assume that a ship and the cable behind it are swept sideways by the component of surface current velocity perpendicular to the ship's heading. An electromotive force equal to $vH_z l$ is induced in the cable, where v is the sideways velocity of the ship and cable, H_z is the vertical component of the earth's magnetic field, and l is the length of the cable. If electrodes making good contact with the sea are fitted at both ends of the cable, and *were there no potential induced in the sea by virtue of its motion*, the electromotive force induced in the cable could be measured by a potentiometer and the velocity determined by the relation expressed above. In general, however, there is a potential gradient component developed in the ocean in the same direction as that in the cable; and since there is no way of knowing exactly what this ocean potential is except by knowing a great deal about the ocean velocity field, the method cannot give true surface velocity. In ocean current systems that are broad compared to the depth of the ocean, the potential gradient developed in the ocean is approximately $\bar{v}H_z$, where $\bar{v}$ is the velocity averaged over the total depth. Since in the Gulf Stream the strongest currents extend to only a fraction of the total depth, $\bar{v} < v$, the towed electrodes give a good approximation to surface velocities, the deficiency varying between 2 and 40 per cent.

Air-borne radiation thermometer.—Stommel and Parson (see Stommel *et al.*, 1953) conducted some field experiments to determine the feasibility of making rapid synoptic surveys of the surface temperature over the Gulf Stream by airplane, using an infrared bolometer. The bolometer is exposed alternately to the sea surface and to a black body in the plane. The chopped radiation, falling on the bolometer, is restricted to the band of maximum water-vapor transmission ($8–13\,\mu$) which contains about 30 per cent of the energy of the ocean's emitted radiation. The airplane is flown below an altitude of 1000 ft., and preliminary tests show that readings accurate to 1° F. are obtainable. The contrast in surface temperature across the Stream is about 4° F. in late summer, 20° F. in late winter. It is estimated that under favorable weather conditions for flying, the plane could make about six zigzag crossings of the Stream from Cape Hatteras to longitude 65° W. in a single 10 hr. flight. Such rapid surveys, which might be made twice a month, are still in the planning stage.

LORAN

Loran is a form of radio navigation developed during the Second World War by the Radiation Laboratory of the Massachusetts Institute of Technology. Its coverage is not world-wide, but fortunately there is a network of stations covering the central sector of the Gulf Stream System. A hyperbola of position is determined by accurately measuring (to 1 μsec.) the difference in time between pulse signals sent at a constant rate from two shore stations which emit in perfect synchronization. A Loran fix is obtained by the intersection of several lines of position. Under the best circumstances the fix is accurate to within 1/4 nautical mile, and fixes can be made successively as often as every 10 min. This is a very great advance over the method of celestial navigation, which is usually limited to obtaining fixes at sunrise and sunset, and a latitude sight at noon.

Chapter Three

THE GEOSTROPHIC
RELATIONSHIP

It is probable that the reader unfamiliar with the literature and conventions of meteorology and oceanography will find that the manner in which the hydrodynamical equations are introduced and abbreviated in this and later chapters is neither sufficiently explained nor fully justified. Because of the specialized nature of the topic of this book it did not seem advisable to me to try to develop the formulation from first principles. The unsatisfied reader is therefore advised to refer to a good treatise on dynamical meteorology, or to Proudman's (1953) textbook on dynamical oceanography, where an adequately detailed exposition can be found. Lamb (1932) gives a concise derivation of the equations of motion referred to a rotating reference frame.

THE BALANCE OF FORCES IN THE GULF STREAM

In the ocean there are two sets of forces which are nearly always almost balanced. The first of the near-balances, and perhaps the more obvious, is that in the vertical between gravity force[1] and the vertical pressure gradient. We introduce a system of rectangular coördinates in the ocean, with the x-axis directed toward the east, the y-axis directed toward the north, and the z-axis directed vertically upward. The expression for this hydrostatic balance of forces is

$$\frac{\partial p}{\partial z} = -g\rho, \tag{1}$$

[1] Gravity is defined as the vector sum of the gravitational force due to the earth's mass and the centrifugal force due to its rotation.

where p is pressure, g is the acceleration due to gravity, and ρ is the density of the ocean water. In general, the order of magnitude of each of these two terms is 10^3 dynes/cm.[3]. Under surface waves, where there are vertical accelerations, terms up to the order of magnitude of 10^2 dynes/cm.[3] may be expected, but these are important only in the upper 100 m. The acceleration term for ocean tides is much less, perhaps of the order of magnitude of 10^{-6} dynes/cm.[3]. The vertical Coriolis force due to horizontal motions is not likely to exceed 10^{-2} dynes/cm.[3] anywhere in the ocean; and, because large-scale vertical motions are so small, this force may be neglected for practical purposes, such as in the computation of depth in terms of measured pressures. The balance of forces expressed by this equation is therefore correct to a very high order of approximation for most large-scale oceanic phenomena.

The horizontal equations of motion for large-scale oceanic features exhibit a similar balance between two large forces compared to which most other terms in the equations of motion are often small. In order to illustrate this, we shall first write the two equations of motion including only these two forces (the so-called geostrophic equations):

$$\rho f v = \frac{\partial p}{\partial x}, \tag{2}$$

$$-\rho f u = \frac{\partial p}{\partial y}. \tag{3}$$

The symbols u and v are the x and y components of velocity, and f is a quantity called the Coriolis parameter $f = 2\omega \sin \phi$, where ω is the angular velocity of the earth and ϕ is the geographic latitude. In mid-latitudes the value of f is approximately 10^{-4}/sec. The maximum current velocities in the Gulf Stream range from 100 to 250 cm./sec., and hence the Coriolis force, acting at right angles to the Stream, is of the order of 10^{-2} dynes/cm.[3]. This Coriolis force is nearly balanced by horizontal pressure gradients due to the density distribution in the ocean, as shown in equations (2) and (3).

It is interesting to compare the order of magnitude of these terms with that of other terms in the equations of motion. If there is a local acceleration of the ocean currents at a particular locality such that a 250 cm./sec. current is diminished to zero in the course of a week, the local acceleration term is of the order of magnitude of 4×10^{-4} dynes/cm.[3].

If the current happens to be flowing in a curvilinear path, with a radius of curvature of $\mathscr{R}$, the inertial terms will be of the order of magnitude of $u^2/\mathscr{R}$. For example, if the radius of curvature of the streamlines is 200 km., and the current velocity is 200 cm./sec., the inertial terms in the equations of motion will be of the order of magnitude of 2×10^{-3} dynes/cm.³. This term sometimes approaches the Coriolis force and is not always negligible, especially in meanders of the Gulf Stream. As will be shown in Chapter VIII, the inertial terms in the direction of flow of the Gulf Stream are quite likely to be important, even though the cross-stream force components are essentially geostrophic.

So little is known about the nature of the turbulent shearing stresses in the ocean that it is dangerous to attempt to estimate their order of magnitude. The Laplacian of the horizontal velocity in the Gulf Stream has been observed to reach values as high as 2×10^{-11}/cm./sec. Information concerning the order of the magnitude of the horizontal-eddy viscosity is scarce. We might take as a conceivable maximum value 5×10^7 cm²./sec. Under these circumstances the maximum viscous force in the Gulf Stream would be of the order of magnitude of 10^{-3} dynes/cm.³. Slightly larger eddy viscosity could make the Stream appreciably nongeostrophic.

Under normal conditions, the stress of the wind on the ocean surface is of the order of 1 dyne/cm.². If this stress is distributed evenly over a layer of water 50 m. deep the effect of wind stress by vertical turbulence is of the order of 2×10^{-4} dynes/cm.³. Thus, all these terms are small compared to the Coriolis force and to the horizontal pressure gradients in the Gulf Stream.

One sees, therefore, that in the horizontal equations of motion there is an approximate balance between the term of the Coriolis force and the term of the horizontal pressure gradient. This relation, often called the geostrophic equation, has been of practical use in estimating the velocities, and, more especially, the transports, of the Gulf Stream. The actual numerical process is somewhat involved, but is essentially based on the equations obtained by cross-differentiation of the geostrophic equations and the hydrostatic equation; elimination of pressure results in the following pair:

$$\frac{\partial(vf\rho)}{\partial z} = -g\frac{\partial \rho}{\partial x}, \tag{4}$$

$$\frac{\partial(uf\rho)}{\partial z} = g\frac{\partial \rho}{\partial y}. \tag{5}$$

Thus, in an ocean current flowing toward the north (positive y-direction), $\partial v/\partial z$ at each level is associated with a decrease of density toward the east at that level (in the Northern Hemisphere), and there is no appreciable horizontal gradient of density in the direction of the stream: $\partial u/\partial z = u = 0$. The density of sea water as a function of depth and position along a vertical section through the ocean thus provides us with a basis for computing the vertical gradient of horizontal velocity normal to the section.

Were the velocity known for any one value of z it would be possible to determine the velocity at all z by numerical integration with respect to z. In this way one calculates the geostrophic currents (often called, in oceanographic parlance, by the unfortunate solecism 'dynamic currents') from the density field. Meteorologists use the same principles for computing winds aloft from observed density fields, but they have the inestimable advantage of knowing the true pressure distribution at the earth's surface, and thus there is no uncertainty in the constant of integration. Everything hinges on a proper determination of the constant of integration. In oceanography we speak of a 'level of no motion', that is, a value of z at which we can believe the velocity is zero, and from which we can carry out the vertical integration of differential equations (4) and (5) for the velocity field. The fact that this all-important level of no motion, or reference level, is still, to all intents and purposes, undetermined, is one of the most disconcerting features of physical oceanography. Some methods for estimating this reference level are given in Sverdrup's textbook (Sverdrup et al., 1942, pp. 456–457). One school of thought simply relies on placing the reference level sufficiently deep to be below the most intense horizontal gradients of density. Thus Iselin's (1940, p. 24) transport calculations of the Gulf Stream are based upon an arbitrary choice of 2000 m. as the reference level, or level of no motion.[2] Were the chosen level significantly lower, say at the bottom, the resulting transports would be at least 50 per cent greater; however, the velocities in the very surface layers of the Gulf Stream would not be changed much. Fortunately, the choice of the reference level has less effect upon velocities of surface water than on those of deep water. It is very important to remember that the 2000 m. reference level is completely arbitrary. Defant (1941) has proposed that the level of no motion coincides with the level of no vertical gradient of geostrophic velocity. Since he has been able to find levels of no vertical gradient in most of the ocean, he has drawn up charts of velocity based on this completely intuitive criterion. Finally, the use of continuity of mass and of conservation of various properties such as salt and heat content has been proposed in special cases in which a section covers all possible entrances and exits for water in a closed arm of the sea. Fuglister and I have tried such calculations. They place a very severe load upon the accuracy of the observations and upon the assumption that there is no time variability or mixing in the structure of the deep water. They involve

[2] The depth of an observation level, or reference level, is sometimes given in terms of hydrostatic pressure rather than in linear measure. The decibar is nearly equivalent to the meter. I have used meters even when referring to works which use the decibar unit. Also, in the past oceanographers usually neglected the transports below the reference level.

very small differences of very large numbers. It is difficult to place much confidence in them.

Thus the choice of the reference level for geostrophic calculations becomes mostly a matter of taste, and we should admit that that is ultimately intolerable. The determination of the level of no motion is not a matter for debate, but for direct measurement—a subject to which I shall return in a polemical section at the end of this book.

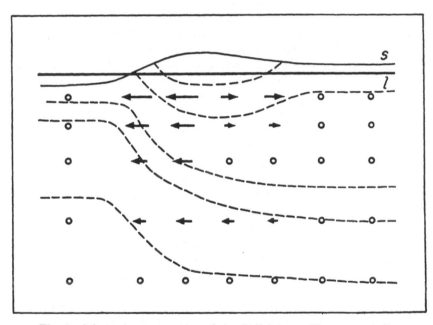

Fig. 1. Schematic cross section of the Gulf Stream. The arrows indicate direction of horizontal pressure gradient. The little circles indicate vanishing horizontal pressure gradient. The current is flowing perpendicular to the plane of the page. The dashed lines are contours of equal water density. The line *l* is a level surface; the line *s* is the actual sea surface.

SCHEMATIC DENSITY AND PRESSURE FIELDS ACROSS THE GULF STREAM

Fig. 1 is a schematic diagram showing a hypothetical cross section of the Gulf Stream. A level surface (such as the ocean at rest would assume) is shown by a heavy solid line. The hypothetical sea surface across the Gulf Stream is shown by the thin solid line. The difference in elevation between one side of the Stream and the other is supposed to be about 1 m. The vertical scale of the diagram is exaggerated very greatly to show the shape of the free surface. The broken lines beneath the surface represent contours of equal density, increasing downward. The spacing of these lines is also vertically exaggerated, but not so much as the free surface. The difference in level of the broken lines is about 700 m. across the Stream.

The diagram extends to a depth of about 1000 m., the lower 3000 m. being omitted. The arrows show the hypothetical magnitude and direction of the horizontal pressure gradient across the Stream. Near the surface the pressure gradient is controlled entirely by the shape of the free surface. At depth, the slope of the density surfaces takes effect until at some depth (in the diagram, 1000 m., but usually assumed[3] to be 2000 m.) the horizontal pressure gradients vanish.

In order that such a distribution of density can long persist, these pressure gradients must be opposed by an opposite and equal force. The essence of the geostrophic method is that we suppose that these pressure gradients are opposed by Coriolis forces acting to the right of the direction of motion of the water. This implies velocities in a direction perpendicular to the plane of the figure, and directed into the page where the arrows point left, and outward where they point right. Since these forces are perpendicular to the direction of motion they neither drive nor brake the motion.

The Gulf Stream is not a river of hot water flowing through the ocean, but a narrow ribbon of high-velocity water acting as a boundary that prevents the warm water on the Sargasso Sea (right-hand) side from overflowing the colder, denser waters on the inshore (left-hand) side.

[3] See footnote 2, on p. 163.

Chapter Four

LARGE-SCALE FEATURES OF THE
NORTH ATLANTIC CIRCULATION

For more than a century the surface features of the oceans have been crudely determined from literally millions of ship reports sent in to the various naval hydrographic services throughout the world. Since no special effort is made to navigate with precision in the deep ocean, the reported surface currents are not determined with very high precision, but the charts drawn up from this large source of data give a good over-all view of surface currents. An example of such a compilation is the United States Navy Hydrographic Office *Current Atlas* (1946).

SURFACE FEATURES

A schematic chart of surface currents is shown in fig. 2. The part of the Atlantic on which our attention is focused in this book is the remarkably intense set of currents on the western side of the ocean, along the open coast of North America. People commonly speak of this whole system of currents as the Gulf Stream, even though very little water from the Gulf of Mexico is actually in the Stream.

Iselin (1936, pp. 73–75) attempted to introduce a well-defined nomenclature for various parts of the current system of the western North Atlantic. The entire set of western currents was to be called the *Gulf Stream System*, and the current from Tortugas, in the Florida Straits, to Cape Hatteras was to be called the *Florida Current*. The old term *Gulf Stream* was retained for the section of the current between Cape Hatteras and the tail of the Grand Banks. Extensions of the Gulf Stream System to the eastward were to be spoken of as the *North Atlantic Current*. The North Atlantic Current, which appears to be made up of a number of separate streams, eddies, or branches (exactly which we do not yet know) is often obscured by a shallow, wind-driven surface movement called the *North Atlantic Drift*, which varies from time to time, depending upon the winds.

In spite of the advantages of this nomenclature, no one has strictly adhered to it. In this book I often use the term Gulf Stream in a more

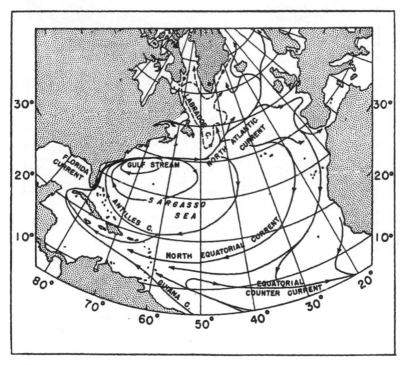

Fig. 2. Chart showing the chief features of the surface-water circulation of the North Atlantic circulation, according to Sverdrup, Johnson, and Fleming (1942, fig. 187). In general, the chart is much oversimplified, and it should be regarded as essentially schematic.

general sense than that proposed by Iselin; and I do not speak of the Florida Current as extending to Cape Hatteras, but restrict the use of this term to mean the current actually within the Florida Straits. Unfortunately, the naming of things is more a matter of common usage than of good sense.

The names of other currents referred to in this book are also shown in fig. 2.

In order to discuss the Gulf Stream in relation to the broad features of the North Atlantic Ocean it is necessary to introduce at this point a number of charts and graphical presentations of data which will be useful in later descriptions and for reference. Thus there are presented, in figs. 3

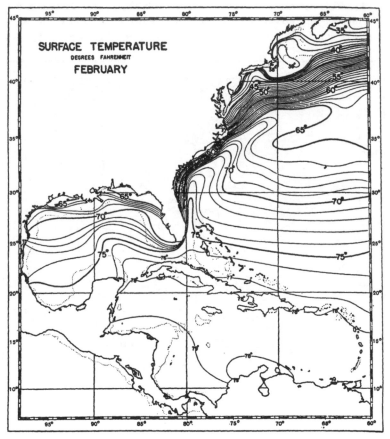

Fig. 3. Contours of surface temperature in the western North Atlantic for February, according to Fuglister (1947, pl. 2).

and 4, the surface-temperature charts prepared by Fuglister (1947, pls. 2 and 8) for the months of February and August. The currents and horizontal temperature gradients are actually much more pronounced than these average charts show, because the averaging processes used in their construction tend to blur the fine and shifting detail of the instantaneous velocity and temperature fields.

It was the close correspondence of the current charts (see fig. 2) and the surface-temperature charts (see figs. 3 and 4) which stimulated the interest of early seafarers in 'thermometric' navigation so many years ago (see Chapter I).

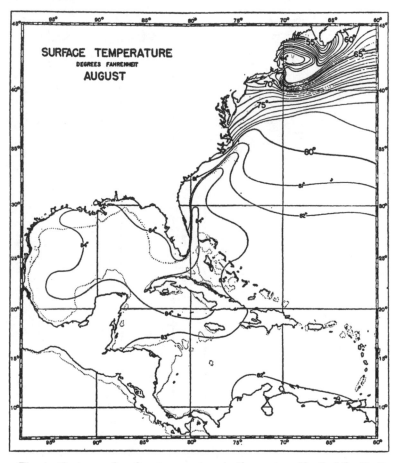

Fig. 4. Contours of surface temperature in the western North Atlantic for August, according to Fuglister (1947, pl. 8).

THE DISTRIBUTION OF PROPERTIES WITH DEPTH

An average temperature sounding (with depth) for two seasons of the year, and on both sides of the Gulf Stream, is shown in fig. 5, *a*. The curve marked 'slope' is the sounding made in the slope water between the continental shelf along the coast and the Gulf Stream near Chesapeake Bay; the curve marked 'central' corresponds to a sounding in the North Atlantic Central

Water on the offshore side of the Stream. Fig. 5, *b*, shows the corresponding values of salinity. One sees that there are three clearly defined thermal regions in the ocean: (i) a surface layer several hundred meters in depth which is subject to seasonal thermal variations; (ii) a zone of marked verti-

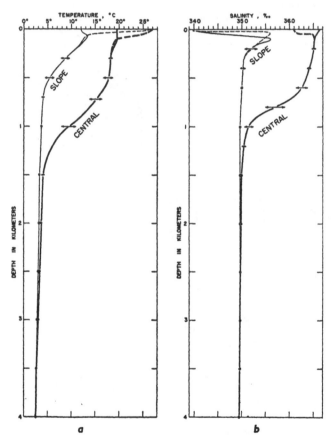

Fig. 5. Temperature and salinity soundings. *a*, temperature soundings at two different seasons of the year in slope and central water. The depth scale is in kilometers. The temperature is in degrees Centigrade. *b*, salinity soundings at two different seasons of the year in slope and central water. The depth scale is in kilometers. The salinity unit ‰ means parts per thousand.

cal temperature gradient, called the main thermocline; and (iii) a very large mass of cold deep water below 1500 m.

Fig. 6 shows the average volume transports of the various surface currents in the North Atlantic Ocean in millions of cubic meters per second according to Iselin (1936). Fig. 7 is a block diagram of the gross features of the North Atlantic Ocean. The observer is standing high above the Gulf of Mexico looking toward the northeast, that is, toward England. The ocean is dissected to show its structure with depth as well as along the surface.

Block 1 represents the westernmost part of the North Atlantic, thus including the Gulf Stream. Block 2 represents the Sargasso Sea. Block 3 represents the area of the North Atlantic Current, its eddies, and its

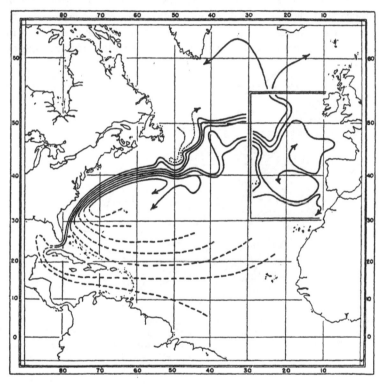

Fig. 6. Iselin's sketch showing sources (broken lines) and pattern (solid lines) of the Gulf Stream System. In the western half of the ocean each transport line represents about 12×10^6 m.³/sec. From Iselin (1936, fig. 48).

multiple currents. The deep water of the ocean, which is indicated by the darkest shading, apparently does not circulate as rapidly as the surface waters. With an average age of some several centuries, it is only slowly renewed by sinking, in very limited areas such as that indicated by the crooked arrow in Block 4. Water between 5 and 16° C. lies mostly in the region of the main thermocline and is shown by intermediate shading in all the blocks. We have no direct information about its sinking rate, but there are some indications that this intermediate water is formed in the

northern half of Block 3, probably mostly in the wintertime. The actual
transports of water at various locations in the Atlantic, as inferred from
geostrophic calculations based on actual data, are discussed in more detail
in Chapter XI. The thickness of the intermediate layer is remarkably constant throughout the whole North Atlantic Ocean; this is in marked contrast to the thickness of the surface layer (water at a temperature higher
than 17° C.), which varies widely from place to place. It seems reasonable
to suppose that the water sinking into this intermediate layer in northern
latitudes (where it reaches the surface in the winter) is eventually mixed
upward into the surface layer in subtropical latitudes. This mixing occurs

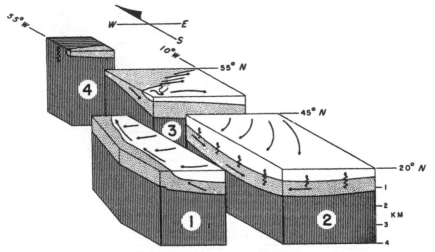

Fig. 7. A dissected-block diagram of the thermal structure and circulation
of the North Atlantic Ocean, as viewed from a great height over the Gulf of
Mexico. The unshaded part is the warmest water; the lightly shaded part indicates the water of the thermocline; and the heavily shaded part represents
cold deep water. The smooth curved arrows indicate the direction of flow of the
horizontal currents of the surface and thermocline; the zigzag arrows indicate
hypothetical slow vertical flows. A description of the individual blocks is
given in the text.

mostly in the fall and winter of the year, when cooling and wind stirring of
the surface layer are at a maximum. To complete the cycle one must suppose that part of the surface water eventually finds its way back into the
northern half of Block 3 through eddies and multiple streams and that
it is there reconverted into intermediate water. A large fraction of the
surface water, however, circulates without transformation in the horizontal wind-driven surface gyre.

Water-mass analysis.—Sverdrup, Johnson, and Fleming (1942) have discussed, in very concise form, the nature and probable origin of various

water masses present in the North Atlantic Ocean. A more comprehensive and detailed study of the area can be found in the earlier monograph by Iselin (1936). Inasmuch as the subject of the present book is limited to the Gulf Stream System itself, it would probably be too much of a digression to try to discuss in detail the features of the entire North Atlantic Ocean. Therefore, the description given here is very brief. The reader who is interested in further detail is referred to Iselin (1936) and to the magnificent oceanographic atlases published in the scientific reports of the

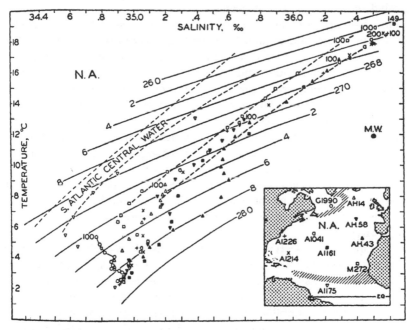

Fig. 8. Plot of temperature versus salinity of selected station data from the North Atlantic, according to Sverdrup, Johnson, and Fleming (1942, fig. 183). A = *Atlantis*; AH = *Armauer Hansen*; G = *General Greene*; M = *Meteor*; N.A. = North Atlantic Central Region; M.W. = Mediterranean Water.

German *Meteor* Expedition. Albert Defant's study (1941) is one of the reports in this series. The two great atlases are Band 5—Atlas (Böhnecke, 1936) and Band 6—Atlas (Wüst and Defant, 1936).

The standard procedure used in water-mass analysis is to plot the hydrographic data from a number of representative soundings on a graph in which the coördinates are temperature and salinity. This graph is customarily called a T-S diagram. Fig. 8, from Sverdrup *et al.* (1942, fig. 183), shows the T-S diagram for a number of specially selected stations in the North Atlantic Ocean. Observations from the upper 100 m. have been omitted because they usually show a great scatter, being variable for different seasons. The depths of the shallowest points in the diagram are indicated by numerals (meters), and the insert chart shows the positions of the various stations.

The T-S diagram is used to identify different kinds of water in the ocean, to note similarities of properties between one region and another, and to obtain a qualitative picture of the amount of mixing between water masses of different geographic location. Four considerations must be kept in mind in the use of T-S diagrams.

First, potential temperature and salinity are conservative properties in the ocean, except in a shallow homogeneous layer near the surface, where evaporation, solar radiation, cooling by back radiation, an exchange of heat with the atmosphere, and similar changes occur. Sea water is compressible; therefore, true temperature is not strictly conservative. The change in temperature of a water parcel that is subjected to an adiabatic vertical displacement of 1000 m. is roughly $0°1$ C., and therefore, for many practical purposes of graphing, the potential temperature and the true temperature on a T-S diagram are interchangeable, although potential temperature is preferable.

Secondly, mixing processes in the interior of the ocean mix both temperature and salinity in the same way. Thus the mixture of two masses of water represented by two points on a T-S diagram lies along a straight line joining them.

The third consideration is that the oceans of tropical and temperate regions are stratified and vertically stable at all depths, with the possible exceptions of the shallow homogeneous surface layer and of certain deep isolated basins on the ocean bottom. Because of the damping of vertical turbulence by the stability, there is a tendency for most major flows to occur along surfaces of equal potential density. The potential density is the density that a sample of water would have if it were brought adiabatically to the surface at atmospheric pressure. It is important to make the distinction between density and potential density, because of the considerable compressibility of sea water. At a pressure corresponding to that present at the average bottom depth of the ocean, 4000 m., water is compressed about 2 per cent. Deep flows tend to occur along surfaces of equal potential density; hence it is important that the potential density be represented on the T-S diagram, to indicate the directions of preferred flow and mixing. The family of smooth curved lines on the T-S plane in fig. 8 represent loci of constant σ_t, a quantity which, unfortunately, does not have any special name, but which is very nearly the same as potential density. Because of its importance to physical oceanography it must be defined carefully.

The density of sea water ρ is always a little greater than unity in the c.g.s. system; therefore, in order to avoid having to write long decimals all the time, the quantity σ has been introduced, and is defined in the following way:

$$\sigma = (\rho - 1)\, 1000. \tag{1}$$

Thus, a density of 1·02721 is written as a σ of 27·21.

The quantity σ_t is defined as the σ of sea water at atmospheric pressure and at the temperature at which it was collected. Therefore, it differs slightly from the σ which would be computed from the potential density of the same specimen, in that no account is taken of the adiabatic temperature change of the specimen during reduction of the pressure to standard pressure. This difference is very small. The essential effect of compressibility of sea water is taken into account by the use of σ_t surfaces on a T-S diagram, instead of the rigorously correct equal-potential-density surfaces.

The relation of density to temperature, salinity, and pressure has been the subject of elaborate and highly precise laboratory measurement. Accurate tables have been prepared. They are discussed in more detail by Sverdrup et al. (1942). A very much abbreviated table for σ_t is given in Appendix III of the present study.

Finally, it is necessary to make one further remark about mixing along a surface of equal potential density. Since mixture on a T-S diagram occurs along straight lines, whereas the lines of equal potential density (or equal σ_t) are curved convexly upward, the mixture of two water masses of the same potential density tends to increase the potential density of the mixture. Although this effect, called cabbeling, should be borne in mind, it is not of great importance in the crude qualitative analysis that is usually done with T-S diagrams.

North Atlantic water masses.—The water in the North Atlantic Ocean is made up essentially of two water masses: one, the so-called *North Atlantic Central Water*, and the other, the *North Atlantic Deep Water*.

The North Atlantic Central Water is that water with a temperature between 8 and 19° C. and a salinity of between 35·10 and 36·70‰. (The symbol ‰ is read 'parts per thousand'.) This water mass is delineated by the two sloping broken lines on the T-S diagram (fig. 8). As can be seen, most of the soundings for temperature higher than 8° lie between these narrow limits. Iselin (1936, figs. 22 and 25) has drawn T-S diagrams for water masses in the western North Atlantic which show even sharper definition, indicating that over much of the Sargasso Sea the salinity of water of a given temperature does not vary much more than the accepted limits of error of salinity determination accommodate. The sounding which lies farthest from the mean in fig. 8 is that for *Atlantis* station

1175, which appears to represent South Atlantic conditions (low average salinity).

The North Atlantic Deep Water is characterized by temperatures between 3·5 and 2°2 C. and salinities between 34·97 and 34·90‰. This body of water composes most of the North Atlantic by volume. It is extremely homogeneous over the entire North Atlantic. It does not take part in the wind-driven surface circulation, but does exhibit large transports in a narrow stream near the western coasts of both the North and the South Atlantic—presumably as a result of the thermodynamically driven part of the circulation (see Chapter XI). It apparently originates in the most northerly parts of the wintertime North Atlantic, but the exact way in which it is formed is something of a mystery (see Worthington, 1954a). Near the very bottom it is probably mixed somewhat with bottom water of Antarctic origin.

Between these two water masses, which may be regarded as the principal masses of the North Atlantic, there are other, smaller, amounts of water which are produced by mixing at intermediate depths along surfaces of equal potential density (σ_t range: 27·2–27·8) with water from the South Atlantic and with an outflow from the Mediterranean Sea. The highly saline water from the Mediterranean is shown by the point M.W. in fig. 8. As the Mediterranean Water leaves the Straits of Gibraltar, it sinks along the 27·6 σ_t surface and mixes with water over most of the eastern North Atlantic. Its presence is obvious in both the two southern *Armauer Hansen* stations. Similarly, there is a mixture at mid-depths with water from the South Atlantic Ocean, called Antarctic Intermediate Water. This water mass, of low salinity, forms at the surface of the South Atlantic in a broad band extending from the Cape of Good Hope to Cape Horn. It then sinks beneath the surface and flows northward; and eventually it crosses the equator and mixes with waters in the North Atlantic. The particular stations chosen for fig. 8 do not show this minimum-salinity water very well except at *Atlantis* station 1175, which is really characteristic of South Atlantic conditions. A very small amount of low-salinity Arctic Intermediate Water is formed in the northern regions of the Atlantic Ocean; the effect of this mass shows up in *General Greene* station 1990.

So far as the study of the Gulf Stream System is concerned, the most important water mass is the North Atlantic Central Water, since it occupies the upper 1000 m. of the central regions of the North Atlantic and takes part in the wind-driven surface circulation which gives rise to the Gulf Stream. In particular, there is a very large mass of nearly homogeneous water at temperatures between 17 and 18° C., as shown in fig. 5, *a*. It is not possible to explain the origin of this water satisfactorily on a quantitative basis. Iselin (1939) calls attention to the fact that most of the σ_t surfaces in the North Atlantic Central Water intersect the ocean surface at points where the surface T-S relation is similar to that at greater depths. He suggests that the surface waters sink along σ_t surfaces, without mixing, preserving the T-S characteristics acquired at the surface. Sverdrup (Sverdrup et al., 1942, p. 145) has suggested that vertical mixing is likely to be of considerable importance in the area. It is possible (see Worthington, 1954a) that the entire treatment of the origin of the North Atlantic Central Water to date has suffered from attempts to imagine it as a stationary phenomenon, in which every water-producing process acts all of the time. Upon closer scrutiny every scheme of circulation that has been suggested exhibits some shortcoming. In addition to the subtleties involved in the thermodynamical processes of a continuously stratified fluid, slow density flows, and imperfectly understood vertical and horizontal turbulent processes, there are a number of constraints of a dynamical sort (p. 122), associated with the wind-driven horizontal circulation of the North Atlantic, which probably play very significant roles in the formation of the North Atlantic Central Water.

A fair idea of the exchange of heat and water across the surface can be computed by semiempirical laws from a knowledge of average air and water temperatures (Jacobs, 1942), but even with this detailed information of net heat energy and water flux at the surface, the problem of the formation of the North Atlantic Central Water defies quantitative explanation.

Water masses of the Gulf Stream.—As is shown in the current chart (fig. 2), most of the water which enters the Gulf Stream System is water previously driven westward by the trade winds. The westward flow is a broad band of moving water, called the North Equatorial Current, which moves slowly, and, especially on its southern side, is very shallow (200 m.). In reaching longitude 60° W. it divides its flow into two parts: one part flows through the Caribbean, in a series of gradually narrowing channels and straits, and eventually finds its way out into the North Atlantic through the Florida Straits; the other, northern, part flows north of the West Indies, then joins the Florida Current over about 8° of latitude. The combined flow, the mature Gulf Stream, leaves the coast at Cape Hatteras. The total transport of the Gulf Stream off Chesapeake Bay (e.g., in April, 1932, 82×10^6 m.^{3}/sec., assuming a 2000 m. reference level) also includes some water which recirculates in a long quasi-elliptical orbit to the southeast of the Stream, but which can hardly be properly called North Equatorial Current water. The transports of these various currents and parts of currents are shown in fig. 6. It is clear that since the various portions of water which make up the Gulf Stream proper come from a wide range of different localities in the North Atlantic, they bring to the Stream rather distinctive T-S relations, and that these may be used, to some extent, to trace the origin of various kinds of water in the Stream. It would be premature to explore this subject further at this point of the book; it is therefore deferred until after the nature of the Stream itself has been discussed in greater detail. A recent section made by Worthington from Nova Scotia to South America along approximately 64° W. longitude, sketched in fig. 9, illustrates how narrow the current of the Gulf Stream is compared to the Equatorial and Caribbean currents that feed it.[1]

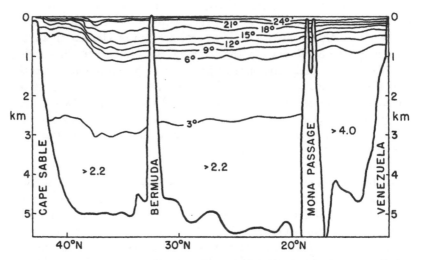

Fig. 9. Vertical section along approximately 64° W. longitude, extending from Nova Scotia, on the left, through Bermuda and then through Mona Passage in the West Indies, and through the Caribbean to the coast of South America. Drawn from deep stations made by Worthington on the *Atlantis* and *Caryn* in 1954. This sketch shows particularly well the contrast between the sharp Gulf Stream, in the northern half of the section, and the broad North Equatorial Current in the southern half. Mr Worthington very kindly let me use some of his observations to construct this figure in advance of his publication of the full details and final interpretation of his work. In deference to his privilege of prior publication this figure is only a sketch. It is not definitive.

The density field of the Atlantic.—Because the three-dimensional density field is fundamental in describing the geostrophic currents of the ocean, and will later be idealized in theoretical models, and also because equal σ_t surfaces are preferred directions for mixing and movement of water masses, it is worth while to reproduce here the charts of the depths of various σ_t surfaces drawn by Montgomery and Pollak (1942) from the *Meteor* data. These are given in figs. 10–15.

[1] In deference to Mr Worthington's privilege of first publication of the data for this new section which he has obtained, I present in fig. 9 only a rough schematic sketch of the section.

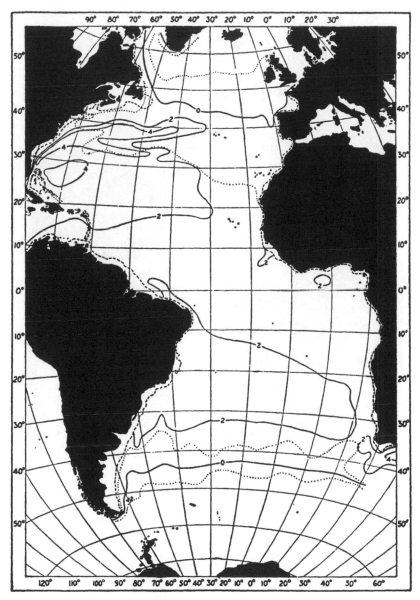

Fig. 10. Depth in hundreds of meters of the 26·5 σ_t surface, according to Montgomery and Pollak (1942, fig. 12).

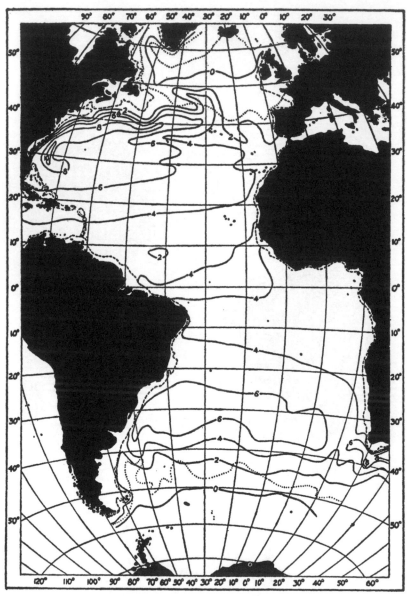

Fig. 11. Depth in hundreds of meters of the 27·0 σ_t surface, according to Montgomery and Pollak (1942, fig. 13).

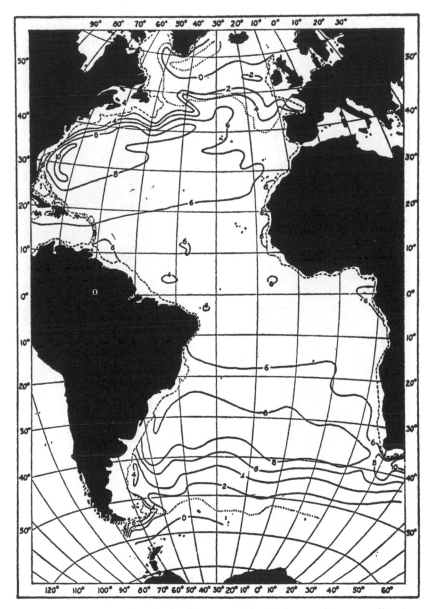

Fig. 12. Depth in hundreds of meters of the 27·2 σ_t surface, according to Montgomery and Pollak (1942, fig. 14.)

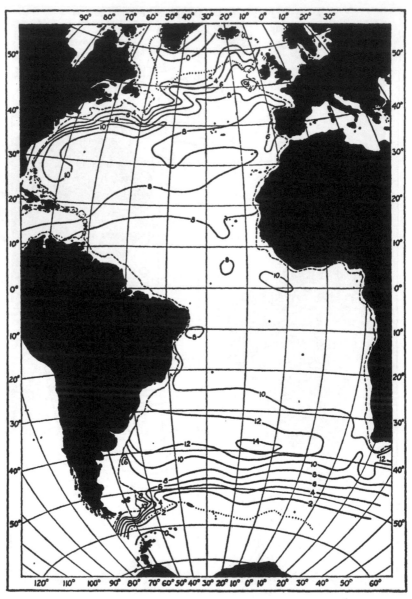

Fig. 13. Depth in hundreds of meters of the 27·4 σ_t surface, according to Montgomery and Pollak (1942, fig. 15).

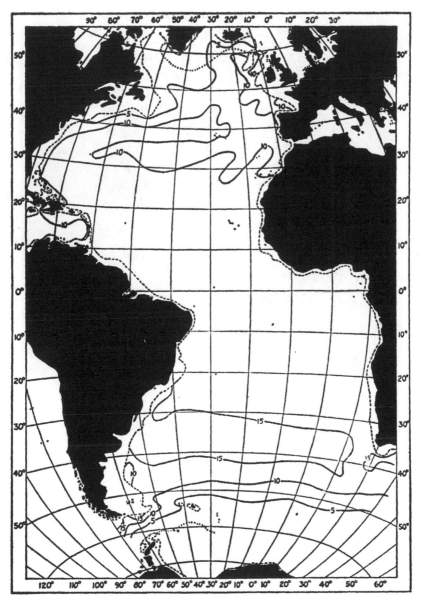

Fig. 14. Depth in hundreds of meters of the 27·6 σ_t surface, according to Montgomery and Pollak (1942, fig. 16).

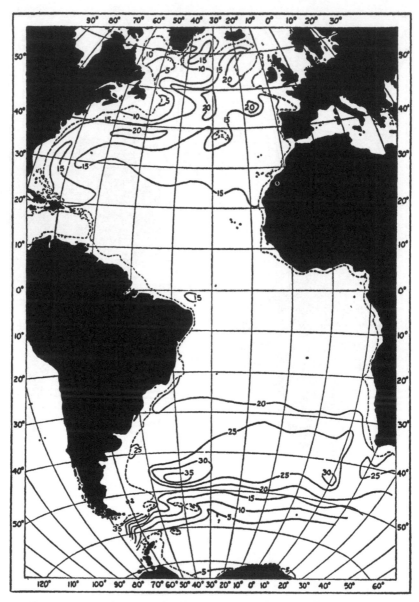

Fig. 15. Depth in hundreds of meters of the 27·8 σ_t surface, according to Montgomery and Pollak (1942, fig. 17).

Bathymetric features along the Stream.—The course of the currents in the Gulf Stream System seems to be determined in part by the submarine topography of the western North Atlantic. Fig. 16 is a bathymetric chart of the region according to Tolstoy (1951, pl. 1). Fig. 17 shows the bathymetry through the Straits of Florida, obtained from Hydrographic Office charts. The depths in these two figures are given in *fathoms* (1 fathom = 6 ft.).

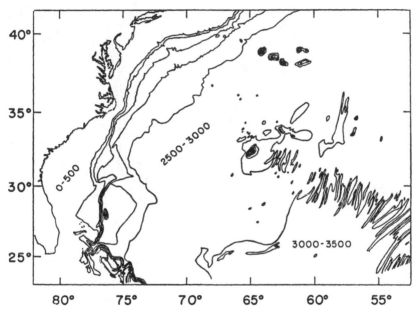

Fig. 16. Bathymetric chart of the western North Atlantic in contours of 500 fathoms, according to Tolstoy (1951, pl. 1). Sea mounts are shown in the upper right quarter of the chart.

The Stream continues on a shelf of about 800 m. depth along the Blake Plateau to about 33° N., where it leaves the shelf. From Cape Hatteras northeast, the Stream flows through a region that is about 4000–5000 m. deep. Since the high-velocity part of the Stream does not penetrate much below 1500 m., it would be difficult to see how bottom topography could influence the Stream here, were it not for the recent discovery (by echo sounder) of numerous sea mounts in the area. Fig. 16 shows the position of some of these.

Farther east the North Atlantic Current encounters the Mid-Atlantic Ridge, which, over large areas, comes to within 2000 m. of the surface.

In order to aid the reader in locating the various sections and positions referred to throughout the text, fig. 18, an index chart has been prepared.

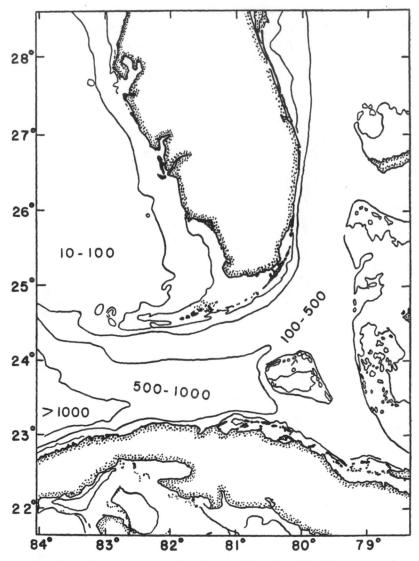

Fig. 17. Bathymetric chart of the Straits of Florida in 500-fathom intervals, but also showing 10- and 100-fathom curves.

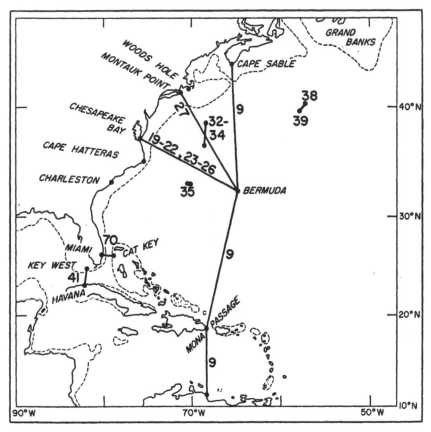

Fig. 18. Index chart, showing the positions of various sections referred to in figures in this book. The numerals on the lines are the figure numbers of the corresponding sections.

Chapter Five

THE HYDROGRAPHY OF
THE GULF STREAM

The first detailed series of hydrographic stations across the Gulf Stream was that begun in 1931 by the *Atlantis*, then the only seagoing research vessel of the newly founded Woods Hole Oceanographic Institution, and consisting of soundings repeated quarterly for a number of years.

'ATLANTIS' SECTIONS OF THE GULF STREAM, 1931-1939

Many early sections were made between Bermuda and Chesapeake Bay (Iselin, 1936) and between Bermuda and Montau' Point, New York (Iselin, 1940). Temperature sections were made at four different seasons, as shown in figs. 19–22; and salinity sections at the same seasons, as shown in figs. 23–26. In these figures the vertical scale above the depth of 2000 m. is much exaggerated (a distortion of 1:370); but below 2000 m. it is less so (1:148). The warmer the water, the less dense. The most striking feature of all these sections is the pronounced change in level of the isotherms in a narrow region. According to the geostrophic relation [Chapter III, equations (4) and (5)], this narrow zone is where the high current velocities, perpendicular to the plane of the page (in figs. 19–26), occur. The surface of the waters to the left of the Stream, called *slope water*, is subject to wider seasonal fluctuations than the Sargasso water to the right of the Stream, where the primary seasonal change is the appearance of a shallow thermocline in the summer.

BATHYTHERMOGRAPH SECTIONS

Fig. 27 shows the kind of detail available in a bathythermograph crossing of the Stream. In discussing the results of a multiple-ship cruise called 'Operation Cabot', Fuglister and Worthington (1951, p. 3) have suggested the following definition of terms:

> Early in the planning and operational stages of Operation Cabot it became evident that precise definitions were needed for

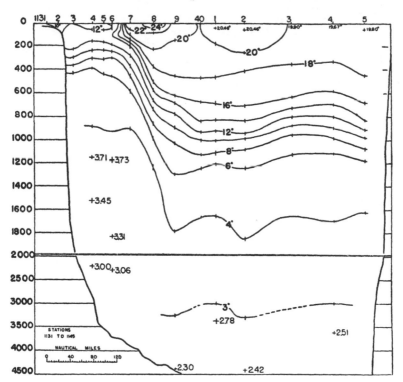

Fig. 19. Temperature section across the Gulf Stream, Chesapeake Bay to Bermuda, February 11–18, 1932, according to Iselin (1936, fig. 3).

the various terms used in association with the Gulf Stream. The frequent references to the 'cold wall', 'edge of the Stream', 'warm core' and 'front' led to a certain amount of confusion and misunderstanding. The term 'inner edge' was most frequently used and most variously interpreted. This confusion is caused primarily because, although the words 'Gulf Stream' denote a current, they also imply a distinct water mass, and secondarily because water masses that may be motionless are included as part of the Stream because they lie below a surface current.

Since the Gulf Stream is a boundary or front in the western North Atlantic between the slope water and the Sargasso Sea we may define it as follows: it is a *continuous* band stretching from the continental shelf off Cape Hatteras to the 50th meridian of

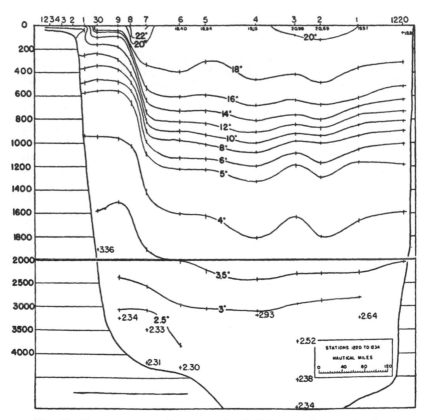

Fig. 20. Temperature section across the Gulf Stream, Chesapeake Bay to Bermuda, April 17–23, 1932, according to Iselin (1936, fig. 5).

longitude, south of the Grand Banks of Newfoundland. This band consists of a pronounced pressure gradient between the warm, highly saline water to the south, and the colder, fresher water to the north. Using this definition then, the inner and outer limits or edges of the Gulf Stream can be defined as the points where this pressure gradient becomes zero. These points can be located only if deep, closely spaced temperature and salinity data are

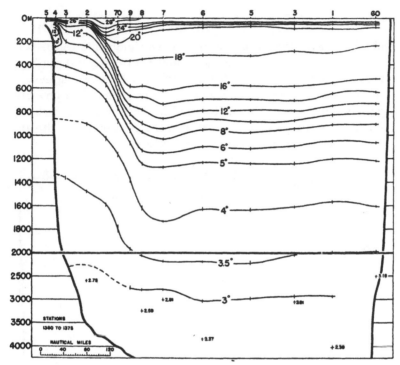

Fig. 21. Temperature section across the Gulf Stream, Chesapeake Bay to Bermuda, August 28–September 3, 1932, according to Iselin (1936, fig. 7).

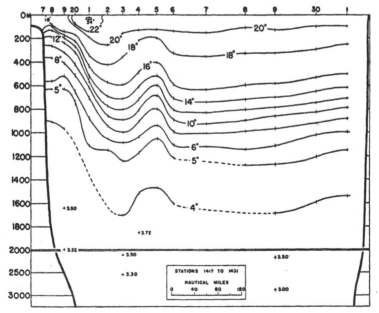

Fig. 22. Temperature section across the Gulf Stream, Chesapeake Bay to Bermuda, November 30–December 5, 1932, according to Iselin (1936, fig. 10).

obtained, and the cross-current pressure gradients calculated.
Also, because of the large eddies found both north and south of
the Stream, any section made across the area must be long enough
to ascertain whether or not more than one pronounced pressure
gradient exists. If only one is found it defines the Gulf Stream,
but if more than one are located then the position of the Stream
cannot be determined by that single section.

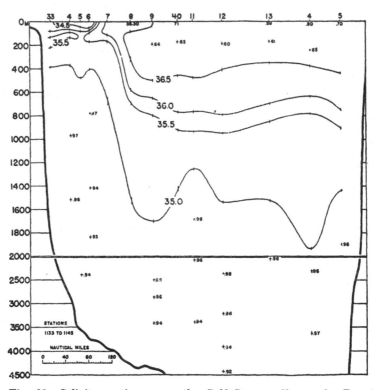

Fig. 23. Salinity section across the Gulf Stream, Chesapeake Bay to
Bermuda, February 11–18, 1932, according to Iselin (1936, fig. 4).

Not to be confused with the inner or left-hand edge of the
Stream is the temperature-salinity boundary at the surface. This
generally abrupt change that occurs to the left of the 'warm core'
may or may not coincide with the left-hand edge of the Gulf
Stream as defined above. This applies also to the color boundary
and the long thick lines of Sargassum frequently seen on the
surface; all of these surface phenomena are apparently associated
with shear zones to the left of the 'warm core' but they are not

necessarily coincident with the left-hand or inner edge of the Gulf
Stream.

The 'warm core' is defined here as that part of the Gulf Stream
containing water warmer than the water at the same depth to the
right, facing down stream, of the current. This 'warm core' is

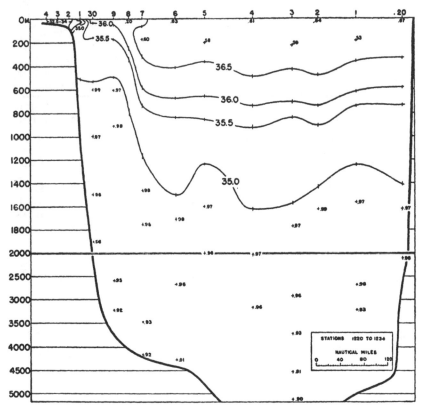

Fig. 24. Salinity section across the Gulf Stream, Chesapeake Bay to
Bermuda, April 17–23, 1932, according to Iselin (1936, fig. 6).

generally 300 to 400 meters deep with the maximum temperature
anomalies at a depth of about 100 meters.

The word 'front' is considered synonymous with the pronounced pressure gradient and therefore with the Gulf Stream
itself.

The term 'cold wall' dates back to 1845 and is still frequently
used to denote the 'inner edge of the Stream' or, according to
Church (1937), 'the temperature gradient between the slope

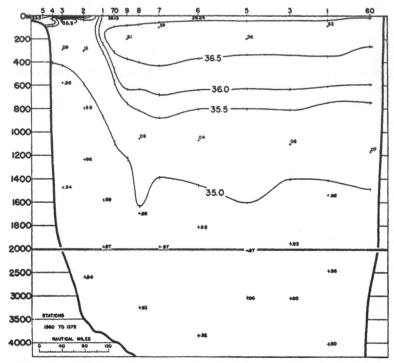

Fig. 25. Salinity section across the Gulf Stream, Chesapeake Bay to Bermuda, August 28–September 3, 1932, according to Iselin (1936, fig. 8).

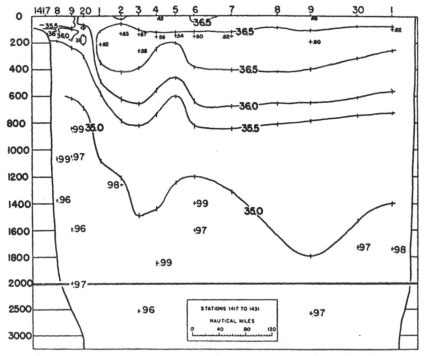

Fig. 26. Salinity section across the Gulf Stream, Chesapeake Bay to Bermuda, November 30–December 5, 1932, according to Iselin (1936, fig. 11).

water and the Gulf Stream'. According to these definitions it could equally well be called the 'warm wall' though in neither case do we have anything resembling a wall. This temperature gradient exists at different depths across the entire width of the Gulf Stream and therefore cannot be considered as something separate or adjoining the Stream. Because of the misleading connotations of the term 'cold wall' it will not be used in this paper.

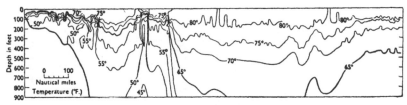

Fig. 27. Sample of temperature sections, in shallow surface layers (upper 900 ft.) across the Gulf Stream, made by means of the bathythermograph. This figure is a simplification of a section published by Iselin and Fuglister (1948, fig. 2). The simplification consists of drawing 5° intervals for the isotherms. Temperatures are in degrees Fahrenheit. The section extends from the continental shelf at the left, across the Gulf Stream, where the 65° F. isotherm drops abruptly, well into the Sargasso Sea on the right. Even in this simplified drawing there is a great deal of fine detail. Just what is the cause of these fine variations we do not know, although internal waves are often mentioned as an explanation. The cold filament of water on the left-hand side of the Stream is clearly shown at a depth of between 200 and 300 ft., just left of center.

MEANDERS AND EDDIES

The position of the Gulf Stream is not always the same, nor is its path even approximately straight. Church (1937) was able to demonstrate the truth of the former statement conclusively on the basis of 1200 thermograph records from ships crossing the Stream. The number of crossings per week was about three or four, and hence it was impossible to develop any detail about the presence of wavelike disturbances along the Stream, or of eddies on either side.

Fuglister and Worthington (1951) have prepared a chart showing the positions of the maximum cross-stream temperature gradients at a depth of 100 m., from all bathythermograph surveys made in the five years 1946 through 1950 (see fig. 28 of the present study). These lines, of course, show a great deal more detail than Church's. They confirm Church's deduction that the position of the Stream varies from time to time, but, more important, they indicate that the Stream does not shift position bodily, but in wavelike patterns which have come to be spoken of as meanders. The first information on the way in which these meanders move was

obtained from the Multiple Ship Survey of 1950. This was the most detailed cruise ever made. Seven ships were employed. The time sequence of
meander patterns was observed.

According to the analysis by Fuglister and Worthington, two meanders
in the western half of the surveyed region moved eastward at a rate of
about 11 nautical miles a day. The water in the swiftest part of the Stream
itself moves more than a hundred miles a day. The amplitude of the
meanders nearly doubled in two weeks. Fig. 29 shows the mean tem-

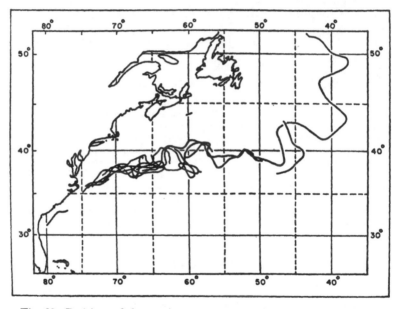

Fig. 28. Positions of the maximum cross-current temperature gradients at
a depth of 100 m., from all surveys in the period 1946–1950, according to
Fuglister and Worthington (1951, fig. 4).

perature, in degrees Fahrenheit, of the upper 200 m. layer of the western
part of the survey area at the beginning of the period of observation.

In the eastern half of the surveyed area a very much distorted meander
was observed to break off into a clearly defined eddy. Fig. 30 shows the
position of this eddy on June 17. The existence of eddies had been inferred
before, but this one was very closely studied because so many ships were
in the area.

Fig. 31 shows a survey of the 'edge' of the Gulf Stream as determined
from a single airplane flight, by means of the air-borne radiation thermometer (Stommel et al., 1953). The black dots indicate the position of
crossing of a strong surface temperature discontinuity as observed from

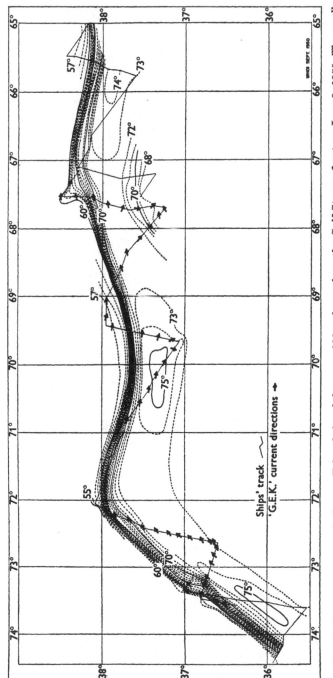

Fig. 29. Mean temperature, in degrees Fahrenheit, of the upper 200 m. layer along the Gulf Stream front on June 8, 1950. The small arrows show current direction as indicated by towed electrodes (Fuglister and Worthington, 1951, fig. 2).

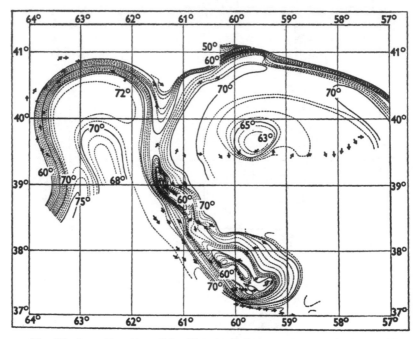

Fig. 30. A continuation of fig. 29 toward the east, showing the large eddy which developed on June 17, 1950 (Fuglister and Worthington, 1951, fig. 7).

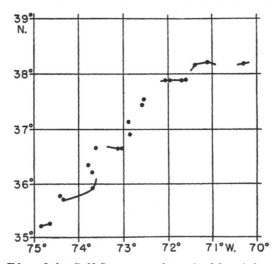

Fig. 31. Edge of the Gulf Stream, as determined by air-borne radiation thermometer (see Stommel et al., 1953, fig. 1). The dots show the position of a strong horizontal temperature contrast. The curved-line segments indicate regions where there was clear visual evidence of the inshore edge of the Stream. The position of this area can be easily ascertained by reference to the index chart, fig. 18.

the airplane. The continuous curves indicate parts of the 'edge' where there was a visible indication of an edge: a sharp change in color, a change in number of whitecaps, and so forth (see frontispiece). The irregularities of the 'edge' in this survey are on a scale different from that of fig. 29.[1]

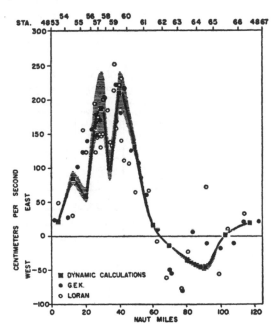

Fig. 32. Closely spaced measurements, by Worthington (1954*b*, fig. 7), of surface current across the Gulf Stream. Surface velocities as computed in three ways are indicated: by geostrophic equation—solid black squares; by towed electrode, with correction applied—solid black circles; by successive Loran fixes—open circles.

VELOCITY DETERMINATIONS ACROSS THE STREAM

Worthington (1954*b*) made three very closely spaced hydrographic traverses across the Stream in October and November, 1950. These are the best ever made, and for this reason it is worth considering one of them in some detail. Fig. 32 shows the surface currents across the Stream as determined by the geostrophic equation, by towed electrodes with a correction factor applied, and by set of the ship as determined by successive Loran

---

[1] A more recent and detailed survey of the variability of the edge of the Gulf Stream is given by W. S. von Arx, D. F. Bumpus, and W. S. Richardson, in *Deep-Sea Research*, 3 (1955): 46–65, under the title 'On the Fine Structure of the Gulf Stream Front'.

fixes. The outstanding features of this velocity profile are the narrowness of the current, the high velocities, and the countercurrent on the right of the Stream. The streaky, banded nature of the Stream, as shown by the curve connecting black squares, is probably not real. Neither the results obtained from the towed electrodes nor those from the Loran fixes show a similar streakiness of velocity across the Stream. It seems likely that the streaki-

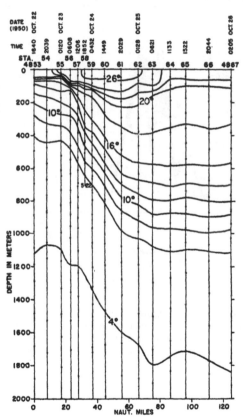

Fig. 33. Closely spaced temperature measurements across the Stream, in degrees Centigrade. By Worthington (1954b, fig. 4).

ness is a false effect arising from internal inertial gravity or tidal waves acting upon the density field.

Fig. 33 shows the thermal field across the same section of the Stream. Only the upper half of the ocean is shown, the total depth in that area being nearly 4000 m. The warm core is wider than the core of high velocity and extends toward the right into a weak countercurrent. This is contrary to the first natural conjecture that high downstream velocities and high

temperature are directly correlated. The main thermocline (10° C. isotherm) drops from 200 to 900 m. across the Stream in less than 70 nautical
miles. The temperature-versus-depth curve of the water below the 16°
isotherm is nearly the same on the two sides of the Stream, except for a
vertical displacement of about 700 m. The chief dissimilarity between the
water above the main thermocline on the one side of the Stream and that

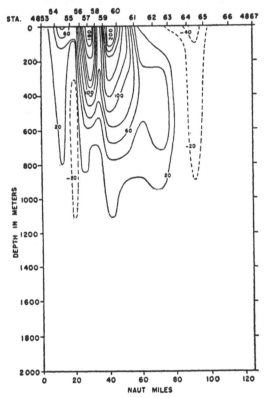

Fig. 34. The geostrophic current velocity (cm./sec.), according to Worthington's (1954 b) closely spaced section (his fig. 10).

on the other consists in differences of temperature of the upper 200 m. and
the presence of a large body of nearly isothermal water (ca. 18° C.) between
200 and 600 m. on the Sargasso Sea (right-hand) side.

From a computation of transport, obtained from fifteen crossings, Iselin
(1940) has shown that the total transport of the Gulf Stream above 2000 m.
in this area is between 76×10^6 and 93×10^6 cm.³/sec. Fig. 34 shows the
geostrophic velocity across the Stream in centimeters per second. The

current is sensible even at a depth of 1000 m. More recent crossings by Worthington, in 1954 and 1955, indicate that the current may be appreciable even at the bottom.

VELOCITY SECTIONS BY DIRECT MEASUREMENT

The only direct measurements of velocity at various depths thus far recorded are those made on a six-day cruise of the *Bear* and the *Caryn* in July, 1952. The measurements were taken in a crest of a meander near Cape

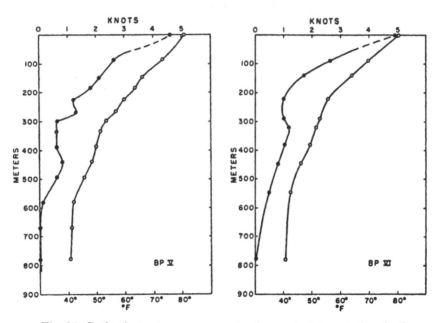

Fig. 35. Bathypitotmeter measurements of current at two stations in the Gulf Stream. The solid black circles indicate velocity; the open circles, temperature. Both series of bathypitotmeter soundings, BP V and BP VI, were made in the left-hand edge of the warm core, in the part of the Stream having the highest velocity, with the ship hove to and drifting with the current, and were obtained by Dr Willem Malkus in June, 1952.

Hatteras. Vertical velocity profiles made by the Malkus bathypitotmeter are shown in fig. 35. The northernmost station was made on the northern edge of the Stream; the other station shown was very close to the center of the Stream. The Watson propeller-type meter was also used and showed little variation of direction with depth. In general, there does not appear to be any great discrepancy between these velocities and the computed geostrophic velocities, but the data are too irregularly spaced for quantitative comparison. Future survey work will, it is hoped, make possible such comparisons.

PROBLEMS OF CONTOURING

Problems of contouring are encountered in two ways: (i) in the construction of vertical temperature and salinity sections; and (ii) in the

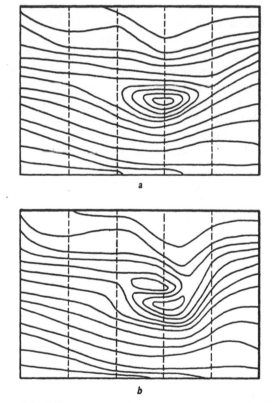

a

b

Fig. 36, *a* and *b*. Different ways of contouring a property the distribution of which is measured only along the vertical dashed lines. See text.

construction of charts showing horizontal distributions. There is never any difficulty in drawing isotherms in a vertical section, for example, when the temperature is a monotonic function of depth, but when the temperature passes through some maxima and minima, the interpretation is ambiguous. Fig. 36 illustrates how a single set of data can be interpreted by contouring in two different ways. The essential decision to make in interpretation is whether there is an isolated central core or whether there is a tongue connected to the lower fluid. If the chart is regarded as a vertical temperature section the likeliest interpretation is that there is no connection and that the cold central mass is fresher; hence, there would be no violation of vertical stability. There is no proof, however, that the water cannot be momentarily unstable (from intense turbulence, the breaking of internal waves, etc.), and hence the alternative interpretation is not automatically ruled out. Simultaneous salinity measurements help to settle the issue, but they are not usually available in bathythermograph sections. Such temperature inversions are common in the side of the Stream toward the coast and have been discussed by Ford, Longard, and Banks (1952), who find them frequently made up of freshened water (30·5–34·5‰ as opposed to 36·5‰ in the Sargasso water in the central North Atlantic).

Similar ambiguities of contouring arise in drawing horizontal charts, and these ambiguities are likely to be even more perplexing because the lines of data on the chart are fewer. It is often impossible to decide whether to draw a stream with an eddy to one side or the other, a sharp S-shaped curve in a single stream, or three separate streams. The only successful and practical way to resolve this difficulty is to meet it at the time when it arises, by proper planning on board the survey vessel. This is a good example to demonstrate why cruises cannot be entirely planned ahead of time and why the uncomfortable job of taking oceanographic data cannot be left entirely to untrained personnel.

THE STRATEGY OF EXPLORATION

In the present state of physical oceanography most of the effort must be devoted to field work, in an attempt to explore and describe the chief physical features of oceanic phenomena such as the Gulf Stream. This does not mean that theoretical studies are not worth while, but it does mean that the chief features are so poorly known that a serious theoretical analysis of the Gulf Stream cannot be made at present. We have seen how the oceanographic surveys of the 1930's have given a broad, general view of the ocean surface circulation and have delineated the major, average thermal and salinity structure of the waters. These early surveys are all characterized by a common strategic practice: the courses and spacing of stations were planned in advance; and the sections actually made corresponded as closely to these preliminary plans as the exigencies of current set, weather, and instrumental failure would permit. The early *Meteor* studies of the South Atlantic, the present-day West Coast survey, and the first ten years of *Atlantis* cruises in the Gulf Stream, are all examples of this type of strategy. So far as the object of exploration is a broad, coarse, climatological-mean picture, these premeditated cruises are successful.

The bathythermograph introduced a new strategy into the game. Its great value lies perhaps less in the continuous nature of its trace with depth than in the fact that it can be used from a ship under way and that the data obtained from it are quickly and easily interpreted. The oceanographer studying the Gulf Stream nowadays keeps a running plot of all bathythermograph information as it comes in. On the basis of his interpretation of this information he frequently alters the course of the ship to explore whatever feature he is particularly interested in. The result is that he directs the ship in a series of traverses which he hopes will cross the Stream again and again, in order to obtain a three-dimensional image of its thermal structure and currents. This new strategy has endless possibilities and ramifications, only a few of which have been explored. For example, the chief tactic has been to follow the Stream by a series of zigzag legs. Fig. 28 shows the positions of the Stream as determined by such zigzag bathythermograph cruises during the period 1946–1950, according to Fuglister and Worthington (1951, fig. 4).

More than a single ship may be employed to advantage in this type of operation: the Multiple Ship Survey of 1950 employed the zigzag tactic and was able to delineate for the first time a meander pattern showing several distinct waves and an eddy breaking off. The expense and effort involved in portraying this picture were considerable.

Another type of strategy was developed during the short 1952 cruise of the *Bear* and *Caryn*; one ship was employed to make the rapid bathythermograph traverses of the Stream, and the other was used in making the slow hydrographic stations and velocity measurements. In this way it was possible to keep a constant check on the position of the slower vessel relative to the chief thermal features of the Stream. This strategy requires at least two ships, one of which should be capable of reasonable speed; they should never be farther apart than a few hundred miles, otherwise the cruise loses the effectiveness of the multiple-ship type of operation. Even two ships face a formidable situation if the ocean feature under observation is changing rapidly. The 1952 *Bear–Caryn* cruise is an example. The evidence of six crossings of the Stream showed that the ships were working on the crest of a meander in which the deep velocity and thermal fields were essentially stationary during the six days spent in the area. The surface salinities, as indicated by a recording conductometric cell, were apparently changing rapidly with time, in the course of a widespread invasion of freshened water from coastal areas (Chesapeake Bay, Delaware Bay, and Hudson River) which was spreading along and across the Stream. The surface salinities plotted in the course of the cruise cannot, therefore, be interpreted as an instantaneous picture at any time. It is possible, of course, to reduce all observations to their approximate positions at any epoch, if the surface velocities are well known and steady, but the practical difficulties are formidable and the final distribution is questionable. Thus the two ships were able to delineate the quasi-stationary thermal and velocity fields fairly well, but were completely unable to do the same for the rapidly changing surface salinity field. It will be very interesting to see what new strategic plan will meet such situations.

There are a number of possibilities. For example, more use could be made of drifting and anchored recording and radiotelemetering instruments. Instruments left on the bottom for long periods might also be used. Electrical potential gradient and pressure seem to be the best variables to measure on the bottom, where fluctuations in temperature and velocity may be insignificant. The airplane can be used in measuring surface temperature rapidly over large areas. It is too early to foresee the new tactics these will demand.

A leg crossing the Gulf Stream may be estimated as 80–100 nautical miles in length. To take twelve hydrographic stations, each of two casts,[2] requires about 70 hr., weather permitting. A leg with a similar number of bathypitotmeter velocity measurements might consume 56 hr. If both hydrographic stations and velocity stations are made, the total time in crossing is 100 hr. These periods may be contrasted to 10 hr. when the bathythermograph alone is used, and 12 hr. for measuring with towed electrodes. An airplane can traverse the same leg in less than 40 min., but the data it can obtain are pretty well limited to surface temperature and photographs.

TRANSFER PROCESSES ACROSS THE STREAM

In 1936 Rossby (1936b) showed that if the Gulf Stream were regarded as a wake stream in the sense used by Tollmien (1926) the observed downstream increase of mass transport could be accounted for. Rossby also suggested that large lateral exchange processes might be at work carrying Sargasso Sea water across the Stream and into the slope water on the left-hand side. It now appears that the increase of transport in the downstream direction can be accounted for by other means (see Chapters VII and VIII), and that the turbulent-jet analogy is not necessary; but the question still remains: what kind of transfer occurs across the Stream?

---

[2] Usually, at any deep-sea hydrographic station not more than twelve reversing bottles and twelve pairs of reversing thermometers are attached to the cable. Three or four separate lowerings must therefore be made, in order to obtain samples at all depths. Each lowering is called a 'cast'.

Indeed, the T-S diagrams (see fig. 37) of water masses below 500 m. on opposite sides of the Stream are so similar that it seems very likely that in deep water such a transfer occurs. The large eddies which detach from the Stream, as exemplified by the eddy observed on the Multiple Ship cruise of June, 1950, certainly effect a transfer of water across the Stream. It should be noted that they are rather solitary phenomena in the Gulf Stream, at least so far as we now know. For example, in June, 1950, there was only one eddy in the process of detachment in a distance of more than 1200 miles along the Stream. The wavelike meanders may transfer momen-

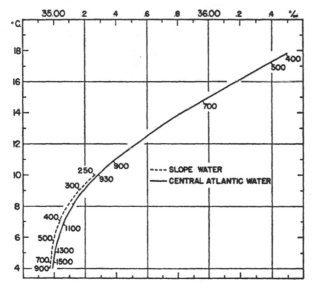

Fig. 37. Comparison between the mean temperature-salinity correlation curve for the slope water and that for the Sargasso water, on the *Atlantis* sections from Chesapeake Bay to Bermuda. From Iselin (1936, fig. 23).

tum, but not heat and salt. A transfer of properties across the Stream on a scale smaller than that of these large eddies does not appear to be indicated. Ford, Longard, and Banks (1952) have called attention to a narrow band of fresh cold water extending from the surface to a depth of about 400 ft., and located within the left-hand side of the Stream. Fig. 38 shows a bathythermograph section through such a filament of cold fresh water. Ford *et al.* point out that of the ninety-seven crossings of the Stream made in June, 1950, fifty-seven show the cold layer. It is quite possible that on the other crossings the bathythermograph lowerings missed the cold layer altogether—it appears to be a very slender ribbon, not more than about 5 miles wide. Fig. 39, from Ford *et al.*, shows the

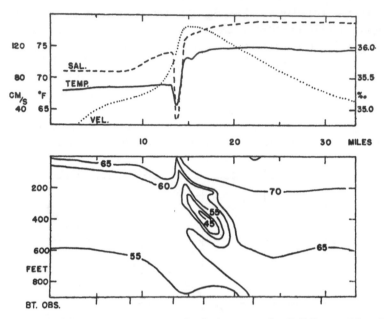

Fig. 38. Salinity, temperature, and velocity across the Gulf Stream (above), showing the fresh filament on the inshore edge; and (below) a bathythermographic section, in degrees Fahrenheit, based on data collected at the same time. From Ford, Longard, and Banks (1952, fig. 3).

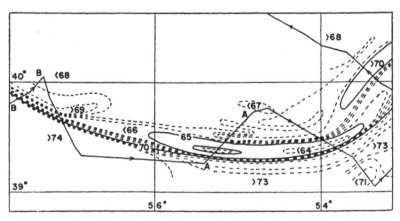

Fig. 39. Chart showing the filament of cold water along the edge of the Stream. The contour lines represent surface temperatures obtained by a continuously recording thermograph on board H.M.C.S. *New Liskeard*. The ship's course is shown by arrowed lines. The current parallels the isotherms. The contours are drawn in by eye, and, of course, are subject to the ambiguities of contouring (see p. 59). From Ford, Longard, and Banks (1952, fig. 2).

position of the cold water in June, 1950. Fig. 40, from the same article, shows the positions of the temperature inversions for the entire Multiple Ship operation, by shaded areas. It is quite evident from the temperature and salinity of this water that it does not come from depth, but must originate from somewhere along the shelf near Cape Hatteras. Were this filament of fresh water a permanent feature, the supply of fresh water required to maintain it would be of the order of magnitude 10^4 m.^{3}/sec.,

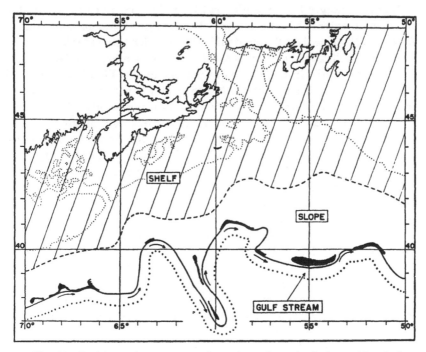

Fig. 40. Positions of temperature inversions along the inshore side of the Gulf Stream, determined on the 1950 Multiple Ship Survey and shown by dark areas. The areas of shelf and slope are indicated. The Gulf Stream current is shown by arrows between the curved solid line and the dotted line, which serve to mark the left- and right-hand sides of the Stream. From Ford, Longard, and Banks (1952, fig. 5).

which could be supplied by river discharge along the coast. The very fact that such a slender filament can preserve its integrity along at least 1200 miles of the Gulf Stream is an indication that small-scale turbulent processes tending to transfer properties across the Stream in the upper layer are inconsiderable.

SOURCES OF THE GULF STREAM WATER MASSES

As was explained in Chapter IV, the water which flows out of the Florida Straits originally comes in large part from the southern half of the North Equatorial Current, and in part from a branch of the South Equatorial Current which is apparently split off from the South Atlantic by the peculiarly wedgelike coastline of Brazil. This water flows through the Caribbean, and then, without mixing with the waters endemic to the Gulf of Mexico, emerges from the Florida Straits in very much the same state as when it entered the Caribbean. Because of the large admixture of Antarctic Intermediate Water at mid-depths which this water has acquired from the South Atlantic, there is a distinct salinity minimum (between 600 and 800 m. depth) in the water coming out of the Florida Straits. This salinity minimum is not so marked in the water flowing westward north of the West Indies. Therefore, the difference in the intensity of the salinity minimum serves as an indicator, or tracer, for distinguishing water in the Stream which has come through the Caribbean from that which has joined the Stream north of the Indies. Iselin (1936) has used the intensity of the salinity minimum in this way to trace the source regions of various portions of the Gulf Stream in various areas of the North Atlantic. He has plotted charts of salinity anomalies which give a rough indication of the sources and degree of mixing of mid-depth waters.

In a detailed analysis of a Gulf Stream section made in April, 1932, off Chesapeake Bay, Iselin (1936) shows that almost all the water colder than 8° C. must have joined the Stream north of the Florida Straits, an amount equal to 15 per cent of the total transport of 82×10^6 m.^{3}/sec. which he computed for the Stream using a 2000 m. reference level. This is consistent with the fact that most of the water at a temperature below 8° C. is blocked from flowing through the Florida Straits by the shallowness of the channel. The water between 8 and 20° C. flowing in the Gulf Stream off Chesapeake Bay is made up of nearly equal parts of water which has come through the Florida Straits and of water from the North Atlantic north of the Indies; together these components amount to about 71 per cent of the total transport. An additional 13 per cent of the total transport occurs in the warm core of water consisting of surface waters warmer than 20° C., but these waters are so subject to wind mixing and to atmospheric cooling that for them the T-S relationship means very little as a tracer. Less than 1 per cent of the total Gulf Stream transport off Chesapeake Bay involves entrained masses of coastal, shelf, and slope waters on the inshore edge of the Stream. The reader is reminded that these figures are based on an assumed reference level.

The T-S method actually gives us a rather broad picture of the possible origins of various parts of the Stream, but in order to refine the method it would be necessary to have some quantitative information about mixing, and it would be necessary to make more widespread and frequent hydrographic sections than are now available. This would require great expense and effort with present techniques. In addition, the uncertainty about the choice of the reference level makes transport calculations rather fuzzy. Thus, in a few words, sparsity of data, ignorance of mixing, and uncertain transport computations conspire to maintain the technique of water-mass analysis at a qualitative rather than a quantitative level.

Although the hope that we shall ever have really frequent and widespread hydrographic data is very dim, there is a need for repeating certain hydrographic stations made in the past. With the waning popularity of the hydrographic station there has been a tendency toward slovenliness in the work which makes some of the new data worthless for water-mass analysis. In order to offset this tendency, as well as to gather new precise data on possible secular changes in the properties of the deep waters, Worthington began, in 1954, to rerun all the old *Atlantis* sections of twenty-odd years ago. This tedious and very demanding work, requiring precision measurements under difficult environmental conditions, is now half finished. When completed, it will be a useful step toward a firm and secure beginning for a long-term study of slow climatic changes of the ocean.

Just as Iselin (1936) used the salinity minimum at the depth of 700 m. to trace water masses in the Gulf Stream, Richards and Redfield (1955) have recently used dissolved-oxygen deficiency to attempt to trace water at about 200 m. It is known that the waters from the Straits of Florida (at 200 m.) are more deficient in dissolved oxygen than are the waters that join the Stream north of the Indies. A diagram analogous to the T-S diagram is used, but the coördinates are dissolved-oxygen concentration and σ_t. The results of Richards and Redfield's analysis of eight available Gulf Stream cross sections made within the period 1950–1953 were highly variable; that is, the amount and position of Florida Straits water at 200 m. present in the Gulf Stream varied greatly from section to section. Of course, eight sections are too few from which to establish the nature of the variability and to relate it to any continuously measured quantity such as variation in difference of sea level across the Straits; but it is certainly worth while to demonstrate the fact of variability itself. The study also showed that there are separate filaments of the Florida Straits type of water in the countercurrent to the right of the current. Possibly this has some bearing on the question whether the countercurrent is frictionally (p. 97) or advectively (p. 123) produced.

THE FLORIDA CURRENT

Most of the water in the Florida Current comes through the Yucatan Channel from the Caribbean Sea, rather than from the Gulf of Mexico (Iselin, 1936). The water is forced through the long channel between the Florida peninsula on one side and the islands of Cuba and the Bahamas on the other by a head of water of about 19 cm., as inferred from a leveling survey across Florida. The Straits between Key West and Havana are 140 km. wide, and their greatest depth is 1500 m. This channel becomes narrower and shallower downstream, takes an abrupt 90° turn to the left, and reaches a minimum cross section off Miami, where the width is 80 km. and the greatest depth is 800 m. The stream then flows north along the continental shelf to about 33° N. latitude, where it leaves the shelf and flows into deep water south of Cape Hatteras. From Cape Canaveral north, the Florida Current increases in mass transport. This is particularly obvious at the place where it leaves the continental shelf and begins to carry water at temperatures lower than 8° C. along with it. There is also supposed to be an increase of volume transport by confluence with the Antilles Current immediately north of the Bahamas (Iselin, 1936), but evidence concerning the transport of the Antilles Current is conflicting.

The mass transport of the Florida Current is estimated to be about 26×10^6 m.^{3}/sec. The computed values vary considerably, as is shown by computations (Montgomery, 1941b) made from four *Atlantis* sections across the Straits at Havana: March 4, 1934; February 19–20, 1935; April 12–13, 1935; and March 24–25, 1938. The transports computed are 26·0, 30·3, 29·0, and $26·0 \times 10^6$ m.^{3}/sec., respectively. There are numerous sources of error in such computations: first, the level of vanishing horizontal pressure gradient must be assumed; secondly, the geostrophic equilibrium may not hold strictly; thirdly, there are gaps between the end stations and the shore where transport must be estimated; and, fourthly, tidal effects on the density field may upset the computation. Woodcock (see Parr, 1937b) made a series of anchor stations across the Straits at Miami which show marked semidiurnal variation in the density structure. Parr (1937b) has described a highly speculative process of cross-stream flow based on Rossby's (1936b) wake-stream analogy. It seems to me that the tidal influence on the density structure in the Straits of Florida might be of the nature of an internal seiche. The dimensions of the channel at Miami and the density structure appear to favor a resonance in a cross-stream semidiurnal internal seiche.

Since the tides in the Gulf of Mexico are very small, one would suppose that there is a tendency for a semidiurnal progressive tidal wave to move upstream from Miami. A study of the tidal constants at various points along the Straits of Florida suggests that this is in fact true, the wave being of the Kelvin type (Thomson, 1871), and the range of tide being greater along the United States coast than along the coast of Cuba or the Bahamas.

Electromagnetic measurement.—An interesting series of observations is being accumulated through the coöperation of the Western Union Telegraph Company and the Woods Hole Oceanographic Institution. Electrodes have been buried in the beaches at Key West and Havana, and connected by a submarine cable. The potential developed by the water

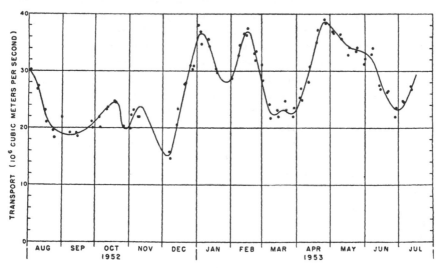

Fig. 41. Mass transport of the Florida Current, as determined from electrical measurements made on the Western Union Telegraph Company cable between Key West and Havana by P. J. Moore and analyzed by Wertheim.

moving through the earth's magnetic field is about 1 volt across the Straits, and, under the assumption of a nonconducting bottom, indicates a volume transport of approximately 26×10^6 m.^{3}/sec. The usefulness and feasibility of this kind of measurement will appear after more years of records have been obtained.

Data for about two years have now been obtained from the submarine cable between Key West and Havana, across the Florida Current. Electrical potential measurements covering several more years will probably be required before any definite conclusions can be drawn concerning fluctuations in the actual volume transport. Wertheim (1954) has examined the first year's readings and has plotted the transport as a function of time (fig. 41). Each point is a 24 hr. average of readings of potential (thus, tidal variations and magnetic storm signals are averaged out). Perhaps the most striking feature of these fluctuations is the extreme rapidity with which major changes in transport can occur.

THE NORTH ATLANTIC CURRENT

There are not a great many observations of the North Atlantic Current (Defant, 1939; Neumann, 1940; Soule, 1950). The *Atlantis* made a section from 37° N. to 52° N. along 30° W. longitude in 1931 (Iselin, 1936), here shown as fig. 42, and the International Gulf Stream Expedition of 1938 (the *Altair* and the *Armauer Hansen*) made some sections in the area. What data there are suggest that the North Atlantic Current is not a single, narrow current like the Gulf Stream off the United States coast, but is made up of several distinct broad currents. Just how permanent these currents are has not been determined. Iselin (1936) and Sverdrup, Johnson, and Fleming (1942) have supposed that the Gulf Stream System splits into a number of branches just east of the tail of the Grand Banks. Fuglister (1951b) has proposed that there is no real branching at all. He has drawn a schematic set of streamlines which fit the data well, and calls his schema the Multiple Current Hypothesis. He suggests that an instantaneous chart would show not a continuous stream, but a number of disconnected filaments of current. This pattern changes from time to time. The possibility that the current system of the North Atlantic Current is irregular and varying and actually discontinuous is very disconcerting. It is even more disturbing to find that Fuglister is able to draw these alternative interpretations even about the water as far west as 65° W. longitude, where we have always thought of the Gulf Stream as unambiguously identifiable. Figs. 43–45 represent three interpretations of the nature of the northeastern parts of the Gulf Stream, according to Fuglister (1955, charts 3a–3c, all based on the same data). These illustrate clearly the difficulties of adequate description of the Gulf Stream System. In my opinion, the interpretation in fig. 45 is a bit forced, from an attempt to spread isotherms apart wherever possible. On the other hand, fig. 43 is forced in the opposite way. Fig. 44 corresponds most nearly to Fuglister's picture of multiple streams and seems intuitively, to me, to be the most natural interpretation.

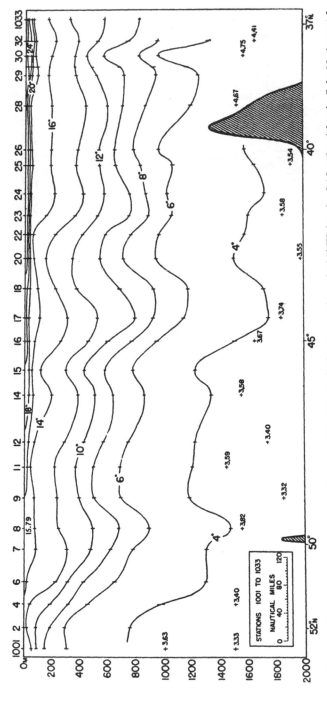

Fig. 42. Temperature section made along the Mid-Atlantic Ridge (approximately 30° W. longitude) by the *Atlantis*, July 26–August 6, 1931, showing a multiplicity of broad currents. From Iselin (1936, fig. 29). The bottom relief is shown by heavy shading. The multiple streams in the North Atlantic Current are clearly shown by the several regions of marked slope of isotherms.

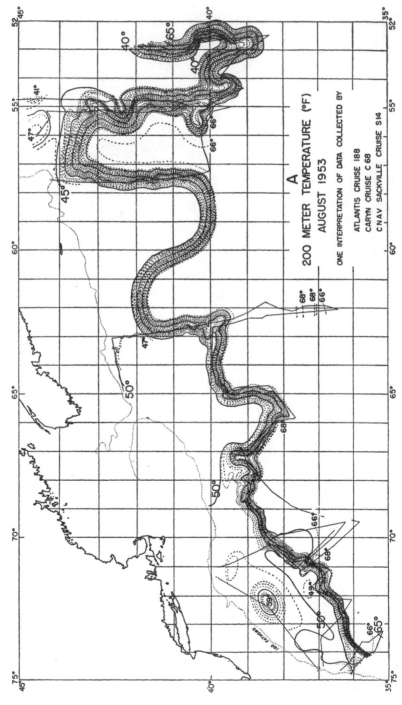

Fig. 43. One interpretation of data collected in August, 1953, by the *Atlantis* Cruise 188, the *Caryn* Cruise C-68, and C.N.A.V. *Sackville* Cruise S-14. The thin straight lines are ship tracks. In this interpretation, A, the Gulf Stream is regarded as a simple, single stream. Compare with alternative interpretations B and C of the same data, presented as figs. 44 and 45 respectively. From Fuglister (1955, chart 3a).

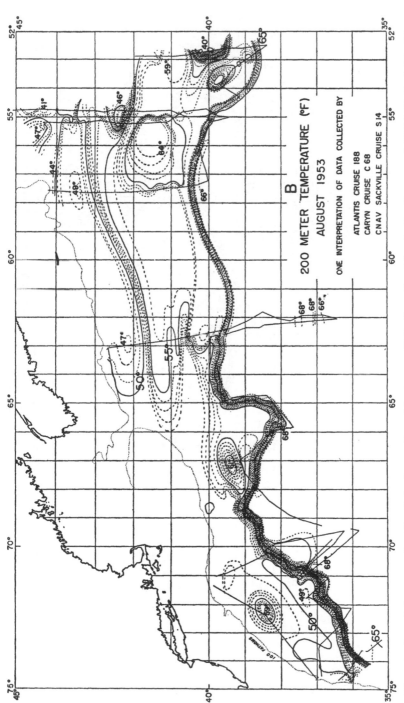

Fig. 44. Second interpretation, B, of the data collected in August, 1953, from the *Atlantis*, the *Caryn*, and C.N.A.V. *Sackville*. Here, the data are regarded as indicating a double stream with slight branching. Compare with interpretations A and C in figs. 43 and 45 respectively. From Fuglister (1955, chart 3 b).

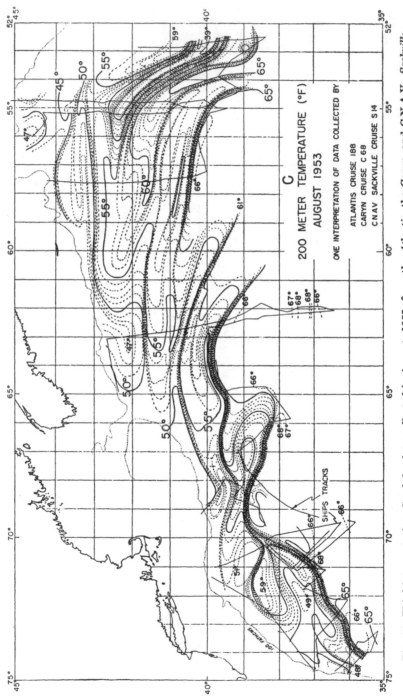

Fig. 45. Third interpretation, C, of the data collected in August, 1953, from the *Atlantis*, the *Caryn*, and C.N.A.V. *Sackville*. Here, the data are regarded as indicating a turbulent confusion of disconnected fragments of the Stream. Compare with interpretations A and B in figs. 43 and 44 respectively. From Fuglister (1955, chart 3c).

Chapter Six

THE WIND SYSTEM OVER THE NORTH ATLANTIC

It is important to describe briefly the winds over the North Atlantic Ocean, because it is generally supposed that the surface circulation of the ocean is caused by them. Many of the older textbooks of meteorology emphasize the mean wind distribution so heavily that one is likely to overlook the considerable fluctuations which occur in wind distribution from day to day. At present, theories such as those described in Chapter VII take into account only the mean wind distribution. Just what the effects of the widespread interruptions and irregularities of atmospheric flow on the ocean circulation are, is still to be told.

NORMAL CIRCULATION

Chase (MS, 1951) has made an extensive study of the surface pressures in the last half-century on the North Atlantic, using the historical weather maps for the Northern Hemisphere (U. S. Weather Bureau, for the period beginning in 1899), and has since incorporated information from the normal weather maps, Northern Hemisphere (U. S. Weather Bureau, 1952). The normal sea-level pressure (yearly average) is shown in fig. 46. By far the largest area of the ocean is covered by a single high-pressure cell with winds traveling in a clockwise direction about its center. The normal sea-level pressures at two different seasons of the year are shown in figs. 47 and 48.

The center of the high is at its most northeasterly position in January; by March it has moved almost 1200 miles to the southwest and is at its

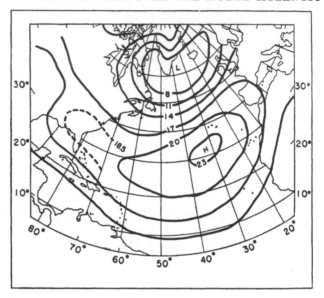

Fig. 46. Map showing annual mean sea-level pressure, according to Joseph Chase (1951, MS). Pressures are given in the excess of millibars over 1000 mb.

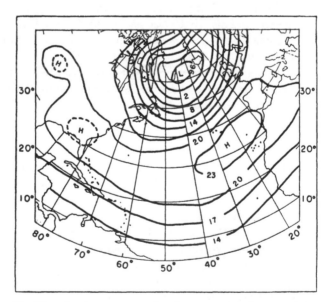

Fig. 47. Map showing mean sea-level pressure for January (Chase, 1951, MS). Pressures are given in the excess of millibars over 1000 mb.

most southerly position. During the spring the center of the high gradient moves due north, and from July on, due east, until in November the center lies almost exactly at the center of the yearly average chart (Chase, 1951). Chase described the changes in pressure distribution as follows (his p. 4):

The general appearance of the mean monthly pressure distributions is similar to that of the annual average. Although the

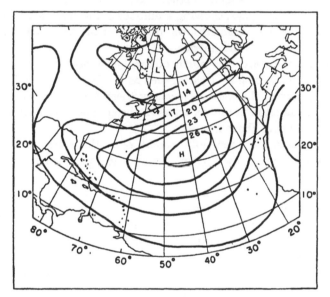

Fig. 48. Map showing mean sea-level pressure for July (Chase, 1951, MS). Pressures are given in the excess of millibars over 1000 mb.

Westerlies are normally stronger in winter than in summer, the circulation around the Bermuda–Azores High, when taken as a whole, is greater in summer. The mean maps for January and July...illustrate this fact. In July the Trades are stronger in proportion to the increased pressure gradient and the east and west ends of the high are much better developed. The Westerlies show a decrease in width of band and in pressure gradient. The decrease is greatest in the area north of 50° N. South of that parallel, the average Westerlies are nearly as strong as in winter. This is the result of more steadiness of direction in summer rather than of high velocities.

INTERRUPTIONS IN THE NORMAL CIRCULATION

Chase (1951, p. 6) also pointed out that only about half of the daily surface-pressure maps look at all like the monthly normal maps: 'A large proportion of the daily maps resemble the average maps in position and orientation of the Bermuda–Azores High and in the circulation about it. These situations are most common in the summer (frequency estimated at some 60 % of the time, and least common in winter [about 30 %]). The frequency for fall is slightly higher than spring.' The rest of the daily maps are quite irregular, showing a number of interruptions which he has classified as: fronts; stagnant lows; linkages of the Bermuda–Azores High to continental highs; and hurricanes.

Some fronts pass to the north of the Bermuda–Azores High without particularly disturbing it. Others, such as the front shown in fig. 49, seem to force it toward the south. Normally these frontal interruptions occur about twice as often in the winter as in the summer, the duration of each interruption being about a week.

The stagnant lows generally persist over the North Atlantic for a longer time. Such lows occur about half as often as the frontal passages.

Frequently a continental high coalesces with the Bermuda–Azores high, causing interruptions of the normal circulation, and these in turn result in widespread irregularities and shifts in wind.

Hurricanes occur in summer and autumn, but primarily in September. They are by far the least frequent interruption to the mean circulation. Fig. 50 represents a very rare situation, that of four hurricanes occurring simultaneously.

THE WIND STRESS ON THE OCEAN

Even though we have a good idea of average wind conditions over the ocean, we can only make crude approximations to the stress produced by these winds on the ocean surface (Rossby, 1936b).

The first approximate values of wind stress were computed (Ekman, 1905) from the mean slope of the water surface in the Baltic Sea, as observed by tide gauges. According to a first-order theory, the local slope of a water surface depends only upon three factors, namely, the surface stress, the bottom stress, and the depth of the water (if the water is assumed to be vertically homogeneous). The vertical internal turbulent-eddy viscosity does not appear in this expression. In a natural body of water where neither the depth nor the wind distribution is uniform, and very little is known concerning the bottom stress, this method is very

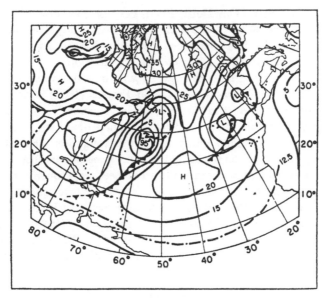

Fig. 49. A daily map showing the disturbance of the Bermuda–Azores high caused by a front pushing in from the northwest (Chase, 1951, MS).

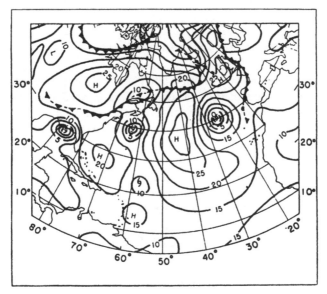

Fig. 50. A rare case of four simultaneous hurricanes over the North Atlantic (Chase, 1951, MS).

uncertain. Recently (Van Dorn, 1953), very precise measurements have been made of the slope of the water surface in a large regular artificial basin. Under such ideal conditions the method gives better results, but the fetch is so short that large waves do not develop.

Experiments to determine the drag of wind on a free water surface have also been made by Keulegan (1951) and Francis (1951) in wind tunnels. There is evidence that the presence of large waves may not have an important effect on the drag coefficient. Francis (*ibid.*) suggests that only the very smallest wavelets and ripples act as roughness elements.

Another method of determining the wind stress stems from laboratory results in the field of aerodynamics. In these experiments it has been found that equations for the stress in a boundary layer can be computed from a knowledge of the variation of wind velocity with altitude over the first 10 m. above the sea. Rossby and Montgomery (1935) applied these aerodynamic equations to observed wind profiles. Table 1, adapted from Sverdrup, Johnson, and Fleming (1942, p. 67), gives drag computed from these equations.

TABLE 1

CORRESPONDENCE OF WIND STRESS (IN DYNES PER SQUARE CENTIMETER) TO WIND VELOCITIES MEASURED FIFTEEN METERS ABOVE SEA LEVEL

Wind velocity (m./sec.)....	2	4	6	8	10	12	14	16	18
Wind stress (dynes/cm.²) ..	0·04	0·16	0·34	1·81	2·83	4·09	5·56	7·25	9·20

Source: Sverdrup, Johnson, and Fleming (1942, table 67).

No account of the thermal instability or stability of the air relative to the water is included in these studies. There is need for ingenious and careful experiments and observations on this problem.

A measure of the wind stress on the sea can be obtained (Sheppard and Omar, 1952) from purely meteorological data, in the following way:

Let u and v be the horizontal components of the wind along axes x and y, where x is in the direction of the surface wind u_0. One then supposes the following equilibrium to hold in the x-direction:

$$fv - \frac{1}{\rho}\frac{\partial p}{\partial x} + \frac{1}{\rho}\frac{\partial \tau_x}{\partial z} = 0. \tag{1}$$

Let us suppose that v_g is the y-component of the geostrophic wind; then

$$\frac{\partial \tau_x}{\partial z} = \rho f(v_g - v) ; \tag{2}$$

and we now integrate this equation from the surface $z=0$ to a height $z=h$, where $\tau_x=0$; hence τ_0, the stress at the surface, is given by

$$\tau_0 = -f \int_0^h \rho(v_g - v)\, dz. \tag{3}$$

The method has been used successfully to measure wind stress over the land by a number of investigators, and over tropical seas by Sheppard and

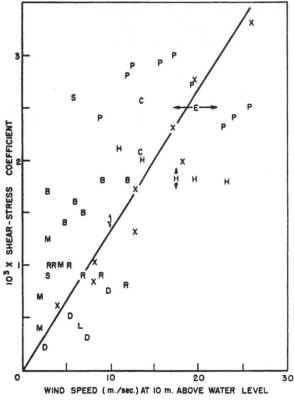

Fig. 51. Plot showing correlation of previous field values of shear stress with model results. X=wind-tunnel results; S=Shoulejkin; P=Palmén; E=Ekman; M=Montgomery; R=Roll; B=Bruch; H=Hela; J=N. K. Johnson; D=Durst; C=Corkan; L=Sutcliffe. Redrawn from Francis (1951, fig. 9).

Omar (1952). Sheppard, Charnock, and Francis (1952) have shown, from a series of observations at the Scilly Islands, that this method does not work well in the westerlies.

Francis (1951) has given an interesting summary of the determination of the drag coefficient γ^2 ($\gamma^2 = \tau_0/\rho_a u^2$, where ρ_a is the density of air, and u is the wind velocity at 6 m.) by these different methods. To date, values of the drag coefficient range from 2×10^{-4} to 10^{-2}. The largest of these were determined in strong winds by the slope-of-the-sea-surface method, the smallest values under light winds from shipboard. Fig. 51 shows various values of the drag coefficient. In spite of the intensive effort that has been made to measure the stress of the wind on the sea, there is so much scatter in the various determinations that the stress is still not well known. It is hard to imagine an object of study more important to the physical oceanography of ocean currents (see Montgomery, 1936b).

Finally, mention should be made of a fundamental difficulty in preparing charts of the mean wind stress over the ocean: because of the non-linear relation between surface wind and surface wind stress, the use of mean wind charts for the computation of mean stress charts is not satisfactory, especially in temperate and polar latitudes, where there are major fluctuations in the observed wind field every few days. The proper way to produce a mean stress chart would be the laborious one of first constructing daily stress charts, and then obtaining vectorial mean stress from them. Actually, the present state of the theory of wind-driven currents does not justify such refinement. An approximate method, proposed by Reid (1948a), is based upon the use of charts of mean wind plus charts of mean variability.

Chapter Seven

LINEAR THEORIES OF
THE GULF STREAM

Perhaps the most striking feature of the large-scale horizontal surface circulation in the North Atlantic Ocean is its east-west asymmetry. Although the mean winds are very broad and diffuse over the entire ocean, the currents along the western shores of the Atlantic are very narrow and intense. This same westward intensification of surface currents is observed in other oceans, and has been the object of study by the writer (Stommel, 1948), who proposed that the latitudinal variation of the Coriolis parameter was the cause of the asymmetry, and by Munk (1950), who first developed a thorough dynamical theory which yields the main features of the ocean circulation and the correct order of magnitude of the transports of ocean currents from the computed wind-stress distribution.

The Kuroshio is the counterpart of the Gulf Stream in the North Pacific. In the Indian Ocean the Agulhas Current hugs the coast of Africa. In the South Atlantic there is the Brazil Current. The coastlines and areas of these oceans are quite different, so we are tempted to think that local bathygraphic peculiarities are not an essential influence in the western intensification of ocean currents; thus the existence of the Straits of Florida is not essential to the formation of the Gulf Stream. If the Antilles were excavated the Gulf Stream would still exist. However, the South Pacific Ocean offers a somewhat embarrassing exception. There does not appear to be any current of great intensity off Australia; in fact, the Humboldt Current off Peru is the strongest South Pacific current, and it lies in the eastern part of the South Pacific. With this important exception the following rule does seem to hold:

On the western edge of most of the world's oceans there is a system of strong currents. Now, we may ask ourselves, why is this so? If the wind system, a broad, widespread phenomenon, drives the ocean currents, how is it that the resulting ocean-current system should be so asymmetrical; in particular, why should the strongest currents be squeezed into a narrow belt on the western edge of the oceans? It is this question which the theories of Stommel (1948), Munk (1950), and Hidaka (1949b) attempt to answer.

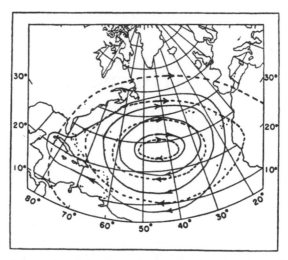

Fig. 52. Schematic wind system (broken lines) and currents (solid lines) which would occur were there no asymmetry in the circulation.

THE THEORY OF WIND-DRIVEN OCEAN CURRENTS

The physical situation is as follows. Suppose we consider a large ocean basin, such as is depicted in fig. 52, with an anticyclonic wind circulating over it. In both figs. 52 and 53 the streamlines of ocean current are shown as solid lines; those of the wind, as broken lines. Our intuition tells us that this wind will produce an ocean surface current in an anticyclonic sense. From what little we know about the distribution of properties in the deeper parts of the ocean it seems reasonable to suppose that the current which the wind induces does not extend to the bottom. Our intuition also tells us that if the ocean-current system has arrived at a state of steady motion under the stress of the wind, the free surface and isopleths of density in the ocean will have adjusted themselves in such a way as to produce the horizontal pressure gradients necessary to balance the Coriolis forces acting upon the moving water. But that is about as far as our intuition takes us, and it gives no hint of the necessary asymmetry in the current system. To progress further it is necessary to consider the processes tending to increase or decrease the vorticity of the ocean water. First we shall consider these processes qualitatively, and then proceed to a mathematical demonstration.

The vorticity of a vertical column of water[1] is taken as positive if counterclockwise, and negative if clockwise. If the spin is measured in its relation to the earth, we speak of relative vorticity; by adding the Coriolis parameter to the relative vorticity we obtain the absolute vorticity of the column. Unlike the velocity distribution, the vorticity distribution in the ocean may

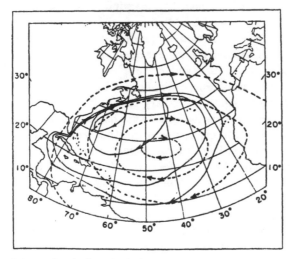

Fig. 53. Schematic relation of wind (broken lines) and current (solid lines), with asymmetry.

be discussed without reference to the pressure field—a fact which greatly simplifies consideration of the qualitative factors. (Formally, the vorticity equation is obtained by elimination of the pressure from the equations of motion by cross-differentiation.) The wind system in fig. 52 is one that would tend to decrease (make negative) the relative vorticity of all the water columns in the ocean. In the steady state of motion which must exist under a prolonged exposure of the ocean to this wind system, the relative vorticity has a fixed value at each point of the ocean independent of time. This

---

[1] The vorticity of an element of fluid is twice its angular velocity. An element of fluid in a shearing motion is subjected to an instantaneous spin. The qualitative reasoning is done entirely in terms of the total vorticity of a vertical water column—that is, the sum of the vertical components of vorticity of all the elements of a vertical water column. The mathematical definitions of several different kinds of vorticity are given in Chapter VIII, equations (3) and (5).

means that other processes must act to counteract the negative-vorticity tendency due to the wind alone.

The linear theories involve two processes, the first of which is friction. Since, as we have already seen, the ocean surface circulation does not seem to be frictionally bound to the bottom, we inquire whether it is frictionally bound to the ocean shores by horizontal eddies. Such a horizontal viscosity would provide a positive-vorticity tendency over the ocean shown in fig. 52. A numerical check, made using (i) values of lateral-eddy viscosity inferred from the distribution of conservative properties, and (ii) a horizontal oceanic circulation which looks like the wind system, without any evidence of asymmetry, requires a circulation many times as fast as the real ocean circulation for lateral friction to produce a positive-vorticity tendency strong enough to balance the wind-stress-vorticity tendency.

The second process is the tendency of planetary vorticity. Regardless of hemisphere, columns of water moving northward without convergence or divergence have a negative-vorticity tendency, and those moving southward have a positive-vorticity tendency. This follows from the conservation of angular momentum, or, to put it in other words, the variation of the Coriolis parameter with latitude. Since the net meridional transport of water across a parallel of latitude is zero (as much water moves north as south), the planetary-vorticity tendency is positive for water in the eastern part of the ocean presented in fig. 52, and negative for water in the western part. Therefore the planetary-vorticity tendency alone is incapable of balancing the wind-stress-vorticity tendency. In the steady state we must have a zero over-all vorticity tendency, by definition. That is, at every point in the ocean the wind-stress-, frictional-, and planetary-vorticity tendencies must cancel out one another.

The distribution of the wind-stress-vorticity tendency may be regarded as fixed, let us say of an order of magnitude -1. If there were only a broad current system without the asymmetry that is actually observed but with a transport of water similar to the observed transport, the frictional-vorticity tendency would be of a smaller order of magnitude, say $+0.1$, and the planetary-vorticity tendency would be of the order -1 in the western and $+1$ in the eastern part of the ocean shown in fig. 52. Thus there would be an approximate balance of tendencies in the eastern part of the ocean, but the western part would not be in equilibrium; hence, for such an ocean as this in a steady state of motion, a symmetrical current system is not physically possible. The state of affairs for a symmetrical circulation is summarized in table 2.

If we let the current system be strongly asymmetrical, as it is in fig. 53, we do not seriously affect the balance between the wind-stress-vorticity tendency and the planetary-vorticity tendency in the eastern part of the

TABLE 2

Vorticity Tendencies in a Symmetrical Circulation

Vorticity tendency	North-flowing currents in western side of ocean	South-flowing currents in eastern side of ocean
Wind-stress	− 1·0	− 1·0
Frictional	+ 0·1	+ 0·1
Planetary	− 1·0	+ 1·0
Total	− 1·9	+ 0·1

ocean, where the frictional-vorticity tendency is very small. However, on the western side, where the velocity and shear are great because of the narrowness of the western meridional current, the frictional- and planetary-vorticity tendencies are greatly enhanced, to orders of magnitude greater than that of the tendency of wind-stress vorticity, say + 10 and − 10 respectively. In this way we may achieve a balance between the tendencies in the western side of the ocean as well, and so obtain a steady state. Table 3 shows the balance of terms in an asymmetric circulation.

TABLE 3

Vorticity Tendencies in an Asymmetric Circulation

Vorticity tendency	North-flowing currents in the western edge	South-flowing currents over remainder of ocean
Wind-stress	− 1·0	− 1·0
Frictional	+ 10·0	+ 0·1
Planetary	− 9·0	+ 0·9
Total	0·0	0·0

If one analyzes ocean currents in this qualitative manner for anticyclonic and cyclonic wind systems in both hemispheres, he will find that the necessary intensification of the ocean circulation is always on the western side of the ocean. This simplified analysis applies only to the viscous theories of the Gulf Stream discussed in this chapter, not to the nonlinear theory presented in Chapter VIII.

WESTWARD INTENSIFICATION

In 1946, when I first began to study the North Atlantic circulation, Dr R. B. Montgomery suggested to me that the marked east-west asymmetry of the surface circulation was one of the main features requiring explanation.

Soon afterward, I proposed a simple model of a wind-driven ocean

(Stommel, 1948), to show the important effect on the transport lines which results from the variation of the Coriolis parameter with latitude.

A rectangular ocean is envisaged, with the origin of a Cartesian-coördinate system at the southwest corner (see fig. 54). The y-axis points northward, the x-axis eastward. The shores of the ocean are at $x=0, r$, and $y=0, b$. The ocean is considered as a homogeneous layer of constant depth D when at rest. When currents occur, as in the real oceans, the depth differs from D everywhere by a small variable amount h. The quantity h is much smaller than D. The total depth of the water column is therefore $D+h$, D being everywhere constant, and h a variable yet to be determined.

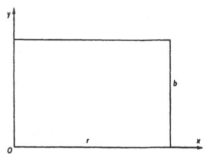

The winds over the ocean are the trades over the equatorial half of the rectangular basin, and prevailing westerlies over the poleward half. An expression for the wind stress acting upon a column of unit horizontal area and depth $D+h$ must include this dependence upon y. A simple functional form of the wind stress is taken as $-F \cos (\pi y/b)$.

Fig. 54. Coördinates and boundaries of the rectangular ocean basin used by Stommel (1948, fig. 1), but with notation changes. The x-axis points toward the east; the y-axis to the north. The dimensions of the rectangular basin are r and b.

To keep the equations of motion as simple as possible, the component frictional forces are taken as $-Ru$ and $-Rv$, where R is the coefficient of friction, and u and v are the x and y components of the velocity vector, respectively. The Coriolis parameter f is also introduced. In general, it is a function of y.

The vertically integrated steady-state equations of motion, with the inertial terms omitted, are written in the form

$$0 = f(D+h)v - F \cos \frac{\pi y}{b} - Ru - g(D+h)\frac{\partial h}{\partial x}, \tag{1}$$

$$0 = -f(D+h)u - Rv - g(D+h)\frac{\partial h}{\partial y}. \tag{2}$$

The quantities u and v are taken to be independent of depth, an assumption which simplifies the analysis, but requires regarding the wind as essentially a body force instead of a surface stress. To these, the equation of continuity must be added:

$$\frac{\partial[(D+h)u]}{\partial x} + \frac{\partial[(D+h)v]}{\partial y} = 0. \tag{3}$$

Cross-differentiation of the first two equations and use of the third result in the following equation:

$$v(D+h)\frac{\partial f}{\partial y}+\frac{F\pi}{b}\sin\frac{\pi y}{b}+R\left(\frac{\partial v}{\partial x}-\frac{\partial u}{\partial y}\right)=0. \tag{4}$$

In the actual oceans, h is so much smaller than D that to a first degree of approximation this may be rewritten as

$$\frac{D}{R}\beta v+\gamma\sin\frac{\pi y}{b}+\frac{\partial v}{\partial x}-\frac{\partial u}{\partial y}=0, \tag{5}$$

where the following definitions have been made: $\beta=\partial f/\partial y$; and $\gamma=F\pi/Rb$. This equation is called the vorticity equation. To the same degree of approximation, the equation of continuity may be replaced by

$$\frac{\partial u}{\partial x}+\frac{\partial v}{\partial y}=0. \tag{6}$$

A stream function ψ is introduced now by the relations: $u=\partial\psi/\partial y$; and $v=-\partial\psi/\partial x$. The vorticity equation is now rewritten in terms of the stream function:

$$\nabla^2\psi+\frac{D}{R}\beta\frac{\partial\psi}{\partial x}=\gamma\sin\frac{\pi y}{b}. \tag{7}$$

The boundary conditions are that the shore of the ocean be a streamline

$$\psi(0,y)=\psi(r,y)=\psi(x,0)=\psi(x,b)=0. \tag{8}$$

If f is a linear function of y, then $(D/R)\beta$ is a constant. The general solution is

$$\psi=XY-\gamma\left(\frac{b}{\pi}\right)^2\sin\frac{\pi y}{b}, \tag{9}$$

where

$$Y=\sum(c_j\sin n_jy+d_j\cos n_jy), \tag{10}$$

$$X=\sum(p_je^{A_jx}+q_je^{B_jx}). \tag{11}$$

The constants A_j and B_j have been defined thus:

$$A_j=-\frac{D\beta}{2R}+\sqrt{\left\{\left(\frac{D\beta}{2R}\right)^2+n_j^2\right\}}, \quad\text{and}\quad B_j=-\frac{D\beta}{2R}-\sqrt{\left\{\left(\frac{D\beta}{2R}\right)^2+n_j^2\right\}}. \tag{12}$$

The quantities c_j, d_j, p_j, q_j are undetermined constants. This solution is very general, but reduces to a simple closed form when the boundary conditions are imposed. First of all, the d_j and c_j vanish, except c_1 corresponding to

$n_1 = \pi/b$. This constant c_1 may be absorbed into p_1 and q_1. When subscripts are dropped, the stream function has the form

$$\psi = \frac{F\pi}{Rb}\left(\frac{b}{\pi}\right)^2 \sin\frac{\pi y}{b}(pe^{Ax} + qe^{Bx} - 1), \tag{13}$$

where

$$p = \frac{1 - e^{Br}}{e^{Ar} - e^{Br}} \quad \text{and} \quad q = 1 - p. \tag{14}$$

The curves ($\psi = $ const.) are the streamlines of the ocean currents.

In order to help visualize the meaning of this solution, it is advisable to compute some numerical examples that will show what role the various parameters play. Three cases are discussed. All involve the same effects of wind stress, bottom friction, and horizontal pressure gradients caused by variations of surface height. The role of the Coriolis force is different in each case. First it is assumed that the Coriolis parameter vanishes everywhere—the case of the nonrotating ocean. Secondly, it is assumed that the Coriolis parameter is constant everywhere—the case of the uniformly rotating ocean. In the third case it is assumed that the Coriolis parameter is a linear function of latitude. Of the three cases, the last one most nearly approximates the state of affairs in the real ocean.

For convenience of the numerical computations the dimensions of the ocean are taken as follows:

$$r = 10^9 \text{ cm.} = 10{,}000 \text{ km.},$$
$$b = 2\pi \times 10^8 \text{ cm.} = 6{,}249 \text{ km.},$$
$$D = 2 \times 10^4 \text{ cm.} = 200 \text{ m.}$$

The maximum wind stress F is taken to be 1 dyne/cm.[2].

The coefficient of friction R is the only quantity for which a value must be devised. If a value of $R = 0.02$ is assumed, the velocities in the resulting systems approach those observed in nature.

The case of the nonrotating ocean.—In the nonrotating ocean the constants p and q are fairly simple. Within 1 per cent, or as closely as graphs may be drawn, p and q are given by

$$p = e^{-\pi r/b}, \quad q = 1. \tag{15}$$

The equation for the stream function is therefore

$$\psi = \frac{F\pi}{Rb}\left(\frac{b}{\pi}\right)^2 \sin\frac{\pi y}{b}[e^{(x-r)\pi/b} + e^{-x\pi/b} - 1]. \tag{16}$$

The east-west and north-south symmetry of the streamlines is immediately evident from this equation. The actual streamlines computed from it are exhibited in fig. 55.

The height contours are computed by integration of the primitive equations and are plotted in fig. 56. The general features of the nonrotating, wind-driven system constitute a broad circulation exhibiting absolutely no tendency toward crowding of the streamlines.

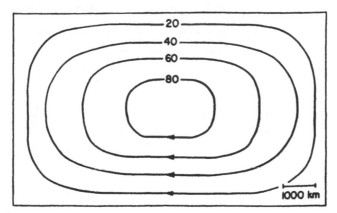

Fig. 55. Streamlines of a nonrotating ocean (Stommel, 1948, fig. 2).

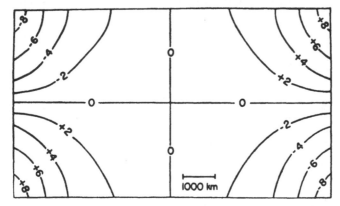

Fig. 56. Contours of surface height in the nonrotating ocean (Stommel, 1948, fig. 3).

The case of the uniformly rotating ocean.—If the Coriolis parameter is a constant 0.25×10^{-4}, the streamline diagram does not differ from that of the nonrotating basin. When the height contours are computed, however, a difference between the two cases becomes apparent, as shown in fig. 57. The large elevation in the central part of the ocean provides horizontal pressure gradients that nearly counterbalance the Coriolis forces. The height contours are not strictly parallel to the streamlines, but nearly so.

Coriolis parameter a function of latitude.—In the real ocean the Coriolis force is a function of latitude. In low latitudes this function is nearly a linear one: $f = \beta y$, $\beta \cong 10^{-13}/$cm. sec. The inequality in the absolute values of the quantities A and B that occurs in this case immediately makes clear the complete lack of east-west symmetry. The streamlines drawn from this formula are shown in fig. 58. The most striking feature of this figure is the

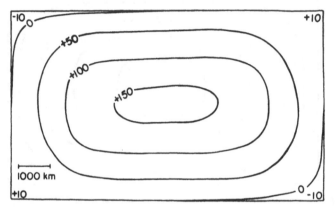

Fig. 57. Contours of surface height in a uniformly rotating ocean (Stommel, 1948, fig. 4).

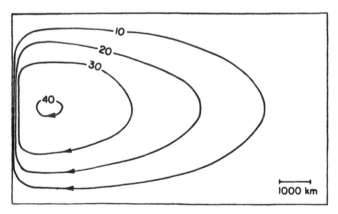

Fig. 58. Streamlines for the case in which the Coriolis parameter is a linear function of latitude (Stommel, 1948, fig. 5).

intense crowding of streamlines toward the western border of the ocean. The rest of the streamline picture is broad and diffuse. The resemblance that the velocity field of this simple case bears to that of the actual Gulf Stream suggests that the westward concentration of streamlines in the wind-driven oceanic circulation is a result of the variation of the Coriolis parameter with latitude.

The height contours computed from this example are shown in fig. 59. To extend the results of this study to the Southern Hemisphere, the reader will notice that since β is unaffected by crossing the equator and F simply changes sign, all the diagrams may be transformed to below the equator by simple reflection across the x-axis. The crowding of streamlines is therefore toward the western border of each ocean, irrespective of hemisphere.

The artificial nature of this theoretical model should be emphasized, particularly the form of the dissipative term.

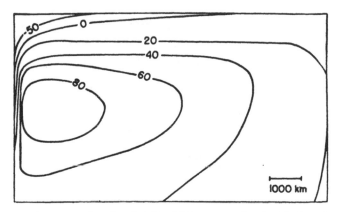

Fig. 59. Contours of surface height which are associated with flow shown in fig. 58 (Stommel, 1948, fig. 6).

MUNK'S THEORY OF THE WIND-DRIVEN OCEAN CIRCULATION

By far the most distinguished theoretical investigation into the wind-driven ocean circulation is that conducted by Munk (1950), who has succeeded in deducing many features of the mean ocean circulation from the wind stress alone. In the preceding section of this chapter we have discussed qualitatively the reason for the great intensification of western currents; in this section we shall follow Munk's quantitative development of the theory.

Let us suppose the ocean surface to be a plane surface at rest: x is eastward, y is northward, and z is upward. The plane $z = 0$ lies at the mean surface; the actual surface is at $z = z_0$.

The equations of steady motion in the horizontal plane are:

$$-\rho f v = -\frac{\partial p}{\partial x} + K_H \left(\frac{\partial^2}{\partial x^2} + \frac{\partial^2}{\partial y^2} \right) u + \frac{\partial}{\partial z} \left(K_V \frac{\partial u}{\partial z} \right), \tag{17}$$

$$\rho f u = -\frac{\partial p}{\partial y} + K_H \left(\frac{\partial^2}{\partial x^2} + \frac{\partial^2}{\partial y^2} \right) v + \frac{\partial}{\partial z} \left(K_V \frac{\partial v}{\partial z} \right), \tag{18}$$

where u and v are the x and y velocity components, p is pressure, f is the Coriolis parameter, ρ is density, and K_H and K_V are the horizontal and vertical coefficients of eddy viscosity. We assume K_H to be constant and uniform.

These equations may be integrated from the surface $z=z_0$ to a depth $z=-h$ beneath which both the currents and the horizontal pressure gradients vanish:

$$P=\int_{-h}^{z_0} p\,dz,\tag{19}$$

$$M_x=\int_{-h}^{z_0} \rho u\,dz,\tag{20}$$

$$M_y=\int_{-h}^{z_0} \rho v\,dz.\tag{21}$$

Now the integrals of the pressure gradients are given as follows:

$$\int_{-h}^{z_0} \frac{\partial p}{\partial x}\,dz=\frac{\partial P}{\partial x}-p(z_0)\frac{\partial z_0}{\partial x},\tag{22}$$

$$\int_{-h}^{z_0} \frac{\partial p}{\partial y}\,dz=\frac{\partial P}{\partial y}-p(z_0)\frac{\partial z_0}{\partial y}.\tag{23}$$

There seems to be little reason to make use of the complete integral of the horizontal shearing-stress terms. We shall assume that

$$\int_{-h}^{z_0} K_H\left(\frac{\partial^2}{\partial x^2}+\frac{\partial^2}{\partial y^2}\right)u\,dz=A\left(\frac{\partial^2}{\partial x^2}+\frac{\partial^2}{\partial y^2}\right)M_x,\tag{24}$$

and that

$$\int_{-h}^{z_0} K_H\left(\frac{\partial^2}{\partial x^2}+\frac{\partial^2}{\partial y^2}\right)v\,dz=A\left(\frac{\partial^2}{\partial x^2}+\frac{\partial^2}{\partial y^2}\right)M_y,\tag{25}$$

where

$$A\cong K_H.$$

The vertical shearing-stress term integrates very simply:

$$\int_{-h}^{z_0} \frac{\partial}{\partial z}\left(K_V\frac{\partial u}{\partial z}\right)dz=\tau_x,\tag{26}$$

$$\int_{-h}^{z_0} \frac{\partial}{\partial z}\left(K_V\frac{\partial u}{\partial z}\right)dz=\tau_y,\tag{27}$$

where τ_x and τ_y are the x and y components of the wind stress applied at the surface. The integrated equations of motion are:

$$-M_yf=-\frac{\partial P}{\partial x}+A\left(\frac{\partial^2}{\partial x^2}+\frac{\partial^2}{\partial y^2}\right)M_x+\tau_x,\tag{28}$$

$$+M_xf=-\frac{\partial P}{\partial y}+A\left(\frac{\partial^2}{\partial x^2}+\frac{\partial^2}{\partial y^2}\right)M_y+\tau_y.\tag{29}$$

We may now introduce a transport function ψ, defined by $M_x = -\partial\psi/\partial y$, $M_y = \partial\psi/\partial x$, and eliminate P by cross-differentiation. We shall assume that $\beta = \partial f/\partial y$ is a constant, and thus obtain a simple linear equation for ψ:

$$\left[A\left(\frac{\partial^4}{\partial x^4} + \frac{2\partial^4}{\partial x^2 \partial y^2} + \frac{\partial^4}{\partial y^4}\right) - \beta\frac{\partial}{\partial x}\right]\psi = \frac{\partial\tau_x}{\partial y} - \frac{\partial\tau_y}{\partial x}. \tag{30}$$

The boundary conditions are that both ψ and its derivative normal to the boundary shall vanish. If the boundaries (coastlines) are taken as forming a simple rectangle $x = 0, r$, and $y = \pm s$, and only an east–west wind-stress system is assumed, $\tau_y = 0$, then an approximate solution is of the form:

$$\psi = rX\beta^{-1}\frac{\partial\tau_x}{\partial y}, \tag{31}$$

where

$$X = -Be^{-(1/2)\,kx}\cos\left(\frac{\sqrt{3}}{2}kx + \frac{\sqrt{3}}{2kr} - \frac{\pi}{6}\right) + 1 - \frac{1}{kr}(kx - e^{-k(r-x)} - 1), \tag{32}$$

where $B = (2/\sqrt{3}) - (\sqrt{3}/kr)$, and $k = \sqrt[3]{(\beta/A)}$.

The ψ-field which Munk (1950, fig. 2) computed for a rectangular basin of Pacific Ocean dimensions, using mean wind data in the fashion described by Reid (1948a), is shown in fig. 60. If we consider mean annual *zonal* winds only, we see that the integrated oceanic wind-driven circulation is divided into closed circulatory systems—or 'gyres', as Munk calls them— bounded at latitudes ϕ_b, where curl$_z$ $\tau = 0$, and with latitudinal axes at latitudes ϕ_a, of extreme values of curl$_z$ τ, which of course do not necessarily correspond to latitudes where $\tau = 0$. The Sargasso Sea in the Atlantic Ocean is thus situated at the inflection point of the mean wind stress between westerlies and Northern Hemisphere trades.

The function $X(x_1)$ along the latitude circles ϕ_a can be interpreted as the total northward transport of the current between $x = 0$ and $x = x_1$, whereas the quantity $X'(x)$ is the transport per unit width. Fig. 61 shows these functions drawn to an arbitrary scale. The region from $x = 0$ to $x = 4/k$ is supposed to correspond to the Gulf Stream.

When X and X' are computed it is found that the equations fall naturally into three parts, each of which dominates in a given sector. At the western edge of the ocean $x \ll r$, and

$$X_w = -\frac{2}{\sqrt{3}}e^{-(1/2)\,kx}\cos\left(\frac{\sqrt{3}}{2}kx - \frac{\pi}{6}\right) + 1, \tag{33}$$

$$\frac{X'_w}{k} = \frac{2}{\sqrt{3}}e^{-(1/2)\,kx}\sin\frac{\sqrt{3}}{2}kx, \tag{34}$$

96

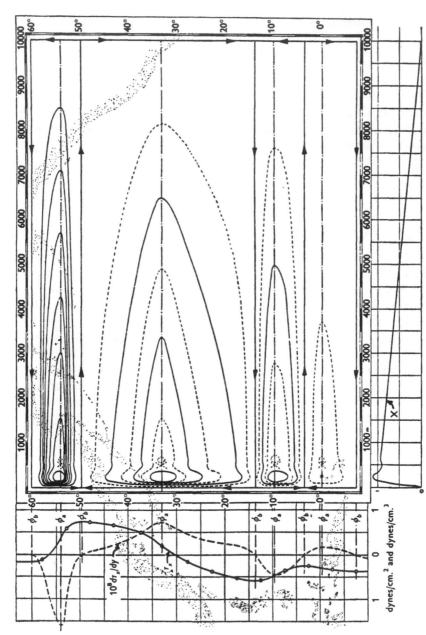

Fig. 60. Volume transport lines $\psi(x, y)$ in a rectangular ocean, as given by Munk for the Pacific. The mean annual zonal wind stress $\tau_x(y)$ over the Pacific and its curl $d\tau_x/dy$ are plotted on the left; the function $X(x)$, on the lower part. The transport between adjacent solid lines is 10^7 m.^{3}/sec. From Munk (1950, fig. 2).

representing slightly *under-damped* oscillations of wavelength

$$L_{\mathrm{w}} = \frac{4\pi}{\sqrt{3}\,k} = \frac{4\pi}{\sqrt{3}\,\sqrt[3]{\beta}} A^{1/3}. \tag{35}$$

Table 4 gives the locations and values of the first few extrema. East of the main current there is a countercurrent of the magnitude $\exp(-\pi/\sqrt{3})$, or 17 per cent of that of the main current. From the analyses of actual data by Wüst (1936) and Iselin (1936) it appears that this theoretical countercurrent is very nearly equal to that observed. In Chapter VIII it is sug-

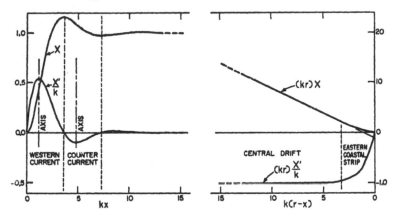

Fig. 61. Plot of equations for X and X'/k giving west–east variation in transport ($\sim X$) and transport velocity ($\sim X'$) from the western shore ($x = 0$) to the eastern shore ($1 - x = 0$). The scale of the central and eastern solutions is exaggerated relative to the scale for the western solution. From Munk (1950, fig. 3).

gested that the Gulf Stream is not essentially a frictional phenomenon. Frictionless models are described, but none give the remarkable similarity to the countercurrent which the Munk theory yields.

TABLE 4

Extrema of X_{W} and X'_{W}

Measurement	Location[a]			
	Western current axis	Western current limit	Counter- current axis	Counter- current limit
x/L_{W}	1/6	3/6	4/6	1
X_{W}	0·45	1·17	1·09	0·97
X'_{W}/k	0·55	0·00	− 0·09	0·00

[a] See fig. 61.
Source: Munk (1950, table 1).

The shape of the X curve for the western current shown in fig. 61 may be compared with the transport function ψ computed geostrophically from hydrographic data for both the Gulf Stream and Kuroshio (fig. 62). The east-west scale of the theoretical curves depends upon the particular choice of k involved, hence of lateral-eddy viscosity. In order to produce a Gulf

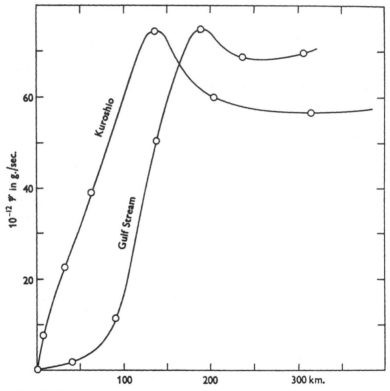

Fig. 62. The mass-transport stream function ψ computed from oceanographic data across the Kuroshio and Gulf Stream. Compare these with the left-hand side of fig. 61. From Munk (1950, fig. 4).

Stream 200 km. wide, Munk chooses $A = 5 \times 10^7$ cm.²/sec. A smaller value of A would give a narrower Gulf Stream.

The more recent cross sections of the Gulf Stream (e.g., fig. 33) suggest that the width of the main current used by Munk is three or four times as great as it should be. Thus, for a purely frictional theory the coefficient of lateral-eddy viscosity must be reduced to about 10^6 cm.²/sec. Although this adjustment of the parameter A is perfectly admissible, there being no independent measure at present, the inertial terms are no longer negligible, and, as we shall see in the next chapter, there is therefore much reason to formulate the boundary layer by using inertial terms rather than friction. For the moment, however, we follow Munk's development.

The total transport of the western current, ψ_{wc}, is independent of A, and, using the value of X_w given in table 4, is simply

$$\psi_{wc} = -1 \cdot 17 r \beta^{-1} \operatorname{curl}_z \tau. \tag{36}$$

Table 5 contains a comparison of the computed transports of various western currents: those computed by Munk (1950, table 2) from the wind stress, and those computed from oceanographic observations.

TABLE 5

THE MASS TRANSPORT OF SOME WESTERN CURRENTS, DETERMINED FROM
THE WIND STRESS AND FROM OCEANOGRAPHIC OBSERVATIONS

Current	Latitude	$10^{12} \beta/$ cm./ sec.	r km.	10^{10} $(d\tau_x/dy)$ g./sec./ sec.	Meridional wind factor	$10^{12} \psi$	
						Computed from wind stress g./sec.	Oceanographic observations g./sec.
Gulf Stream ..	35° N.	1·9	6,500	70	1·30	36	74
Kuroshio Ext..	35° N.	1·9	10,000	50	1·25	39	65
Oyashio Cur...	50° N.	1·5	5,500	− 15	—	− 6·5	− 7
Brazil Cur.....	20° S.	2·2	5,500	− 20	—	− 5·8	− 5 to − 10

Source: Munk (1950, table 2).

The effect of the meridional wind component, which supports the zonal winds, is introduced by a meridional wind factor

$$-\overline{\operatorname{curl}_z \tau} \left(\frac{\overline{d\tau_x}}{dy} \right)^{-1}. \tag{37}$$

The computed transports of the Gulf Stream and Kuroshio are too small by about a factor of two. For example, Munk's computed value of the transport of the Gulf Stream, using $\beta = 1 \cdot 9 \times 10^{-3}$/cm./sec., $r = 6500$ km., and $d\tau_x/dy = 70 \times 10^{-10}$ g./sec./sec., is 36×10^{12} g./sec., as compared to an actual value of about 74×10^{12} g./sec., if the reference level be taken as the depth of 2000 m.

The source of this discrepancy is not to be sought in the physics of the western current itself. The results shown in table 5 apply equally well to the nonlinear theories set forth in Chapter VIII. Munk (1950, p. 92) ascribes the discrepancy to an underestimate of wind stress for low wind speeds:

Finally we may examine the question of why the computed transports of the Gulf Stream and the Kuroshio current amount to only about one half the observed values [see table 5]. It does not seem reasonable that the oceanographic observations should be off by more than, say, 20 per cent, nor that the theoretical expression...for the transport should account for the discrepancy, since it is almost independent of the eddy viscosity and the shape of the ocean basin.

Maury and others have ascribed the North Atlantic circulation, in particular the Gulf Stream, to differential heating between equator and pole, to the freezing of ice, and to other processes that make up the *thermohaline* circulation. If we assume that the circulation were half wind-driven, half thermohaline, it would be a strange coincidence that the general *pattern* of the circulation, such as the boundaries of the gyres, should conform so closely to the general atmospheric circulation. Furthermore, Fuglister [Munk refers to an unpublished manuscript, later published, 1951 a] has found a high correlation between *variations* in the current with variations in the wind. It should also be noted that the thermohaline circulation insofar as it is related to the outflow of river water along the Atlantic seaboard would tend to reduce rather than to strengthen the Gulf Stream. It would seem therefore that the subtropical gyre, and probably also the subpolar gyre, are predominantly wind-driven.

Methods for computing wind stress from the observed wind speeds according to the equation

$$\tau = C_D \rho_{\text{air}} U^2 \qquad\qquad [38]$$

are discussed by Reid [1948a]...Underestimates of τ may first of all result from underestimates in the wind speeds on the climatological charts from which the appropriate averages were taken. The preponderance of coastal stations, and the tendency of ships at sea to avoid regions of high wind, would lead to consistent errors, but these cannot account for more than a fraction of the discrepancies. The weakest link is the drag coefficient C_D which is based on measurements of Baltic storm tides, and a few other measurements...In accordance with the views presently accepted we have assumed $C_D = \cdot 0026$ at high wind speeds, $C_D \approx \cdot 008$ at low speeds, with the discontinuity occurring at Beaufort 4 [Munk, 1947]. In the trade-wind belt of the eastern Pacific, where the winds are predominantly Beaufort 4 and above, Sverdrup [1947] and Reid [1948a] have obtained satisfactory agreement between computed and observed transports. Supposing a value of $\cdot 0026$ were applicable at all wind speeds, then the effect of the south-westerly winds over the eastern Atlantic would make itself felt in a much higher meridional wind factor [table 5], and it can be demonstrated that the discrepancy between computed and observed transports could be largely accounted for. We are therefore led to propose a higher value of C_D at low wind speeds. The transports of the Gulf Stream and Kuroshio current are, after all, probably as good an indicator of the overall stress exerted by the winds on the ocean as any of the measurements on which the value of C_D is now based.

There have been several attempts to obtain better agreement between computed and observed transports. I have expressed my own views in Chapter XI. Hidaka (1949 a) has performed the analysis of the linear theory of wind-driven ocean currents in many different forms. In one study, using spherical coördinates, he has obtained a value of transport of the Kuroshio more nearly equal to that observed than has Munk. It is difficult to see how a change of coördinates can make so great a difference in the transport. Recently, Sarkisyan (1954) has carried out a numerical study dynamically similar to Munk's, but in an ocean shaped very much like the real North Atlantic. He obtained transports for the Gulf Stream of between 70 and 90×10^{12} g./sec., but as I do not know the details of the wind-stress distribution which he used, the significance of his close agreement is not clear to me.

Owing to the existence of a biharmonic operator in the viscous term of the governing equation, there is an exponentially decaying line of vortices to the east of the countercurrent, centered along the axis ϕ_a (fig. 63). There is no good evidence that such a line of vortices actually exists. By some stretch of the imagination one might find confirmation in the charts of Felber (1934) and Defant (1941), but to my mind this is rather special pleading.

Beyond the vortices, over most of the central regions of the ocean the solution reduces to

$$(39) \qquad X_c = 1 - \frac{x}{r}; \quad \frac{X_c'}{k} = -\frac{1}{kr}. \qquad (40)$$

This corresponds, for zonal winds, to the solution of the equation given by Sverdrup (1947, p. 322):

$$\frac{\partial \psi}{\partial x} = \beta^{-1} \operatorname{curl}_z \vec{\tau}, \qquad (41)$$

which does not contain lateral-stress terms. This important relation is discussed further in Chapter XI.

Munk and Carrier (1950) have also investigated the effects of variously shaped ocean basins—in particular, a triangular one, which fits the North

Pacific Ocean more properly than the rectangular model. Munk, Groves, and Carrier (1950) have studied the effect of the nonlinear inertial terms, *assuming they are small*, using a method of successive approximations, and using Reid's (1948b) model of vertical density structure, which consists of an exponential decrease of density upward to the thermocline, and a homogeneous upper layer. They found a slight downstream displacement of the region of maximum currents. The two important observed features

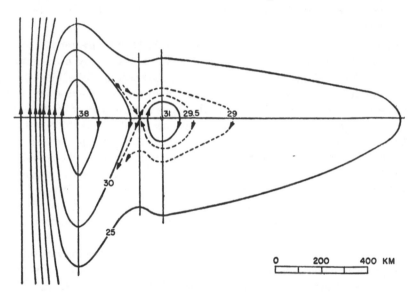

Fig. 63. The volume-transport stream function, in units of 10^6 m.³/sec., near the western boundary for mean annual zonal winds over the Atlantic. The center line is at 31° N. For comparison with the Sargasso Sea circulation, the figure should be distorted by maintaining the west–east orientation of the x-axis and rotating the y-axis clockwise until it coincides with the coast. From Munk (1950, fig. 5).

which were not accounted for in the linear theory, (i) the inshore counter-current and (ii) the continued sharpness of the Gulf Stream long after it leaves the coast, were not found in the higher-order solutions. Sub-sequently, Miyazaki (1952) found that an inshore countercurrent (i) can be obtained formally by letting the coefficient of eddy viscosity decrease toward the coast, and, as we shall see in the next chapter, there is no difficulty in obtaining feature ii if the inertial terms are large enough.

So much of the quantitative aspects of the theory of wind-driven ocean currents depends upon the actual pattern of wind stress used that it is advisable to reproduce for reference the actual distribution of zonal wind stress which Munk has adopted (fig. 64).

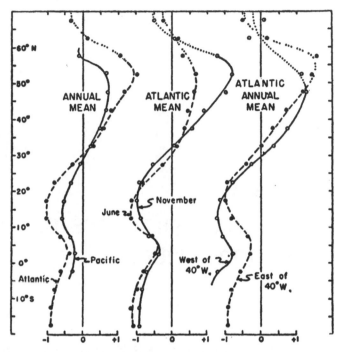

Fig. 64. Distribution of zonal wind stress with latitude. (Zonal wind stress τ_x in dynes/cm.^{2}.) Note the longitudinal asymmetry over the North Atlantic. In the region south of Greenland, where observations are scarce, the stress averages have been corrected for this asymmetry. Dotted lines indicate curves plotted from scanty data. From Munk (1950, fig. 9).

Chapter Eight

NONLINEAR THEORIES OF
THE GULF STREAM

This chapter records attempts to include the effect of the nonlinear inertial terms in theoretical explanations of the Gulf Stream. From a historical point of view it is important that mention be made of the wake-stream theory advanced by Rossby. Rossby's theory of the Gulf Stream assigns a very important role to lateral friction, just as the linear theories discussed in the preceding chapter do, but does not take into account the effect of the variation of the Coriolis parameter with latitude. Next, we shall consider some of the more recent nonlinear theories in which lateral friction is assumed to play only a minor role. We therefore reëxamine the question how large lateral mixing and viscosity really are, and we find that although a definite answer cannot be given, there are grounds for supposing that lateral friction may actually be small as compared to inertial terms.

We shall derive a principle of conservation of a quantity called 'potential vorticity' in order to incorporate the inertial terms in a convenient manner, and then make a vorticity analysis of a Gulf Stream section. After a few preliminary remarks about certain very simple potential-vorticity models, we shall proceed to a study of the recent 'inertial boundary-layer' theories of Morgan and Charney. Finally, reference will be made to Rossby's concept of 'critical' flow as applied to the left-hand edge of the Gulf Stream.

ROSSBY'S WAKE-STREAM THEORY

Fourteen years before the development of the linear theories discussed in the preceding chapter, Rossby (1936a) worked out an interesting nonlinear model of the Gulf Stream in which large-scale lateral mixing was involved. The geostrophic relationship alone does not prescribe a transverse velocity profile for the Stream; and hence Rossby made use of the ideas of a turbulent jet stream from the field of experimental fluid mechanics to define the profile. The Straits of Florida were supposed by Rossby to act as a nozzle, or jet; the Gulf Stream was therefore regarded as a purely inertial stream, interacting with its environment by mixing.

Rossby reduced the problem to one on a nonrotating system, by extracting from the equations the Coriolis-force terms and their associated pressure gradients, and by assuming that the motion is essentially nondivergent. The problem was thus reduced to the one that had already been discussed by Tollmien (1926), who was able to obtain good agreement with experimental results by assuming (i) that the pressure gradient along the jet is small, and therefore that the total momentum transport of the jet is the same at all sections downstream; and (ii) that the shearing stresses acting upon the jet may be described in terms of a mixing length proportional to the distance from the nozzle.

Tollmien found that the mass transport and width of the jet increase downstream. This requires that there be an inflow of water from the surrounding medium into the jet.

Although the mass transport of the Gulf Stream does increase after leaving the Florida Straits, it does not increase after passing Cape Hatteras. The angular spread of experimentally produced jets (Förthmann, 1934) varies greatly, but Peters and Bicknell (1936) obtained spreads of between 8 and 14°. The spread of the actual Gulf Stream in its instantaneous form is less than 1°, although the mean spread averaged over many different sets of observations is greater (Stommel, 1951).

After discussing the analogy to Tollmien's jet in a homogeneous ocean, Rossby (1936a) made some qualitative studies of the effect of the stratification of the sea, by considering an ocean made up of two layers of slightly different density, the lower layer being at rest. By the geostrophic relationship, the difference in level of the interface at the two sides of the jet stream is a measure of the total mass transport of the stream; therefore, if the mass transport increases downstream, it soon becomes impossible to join the interface inside the stream to the level of the undisturbed water masses outside the stream, unless the interface is bent back (up on the right, down on the left) to these levels. This requires, geostrophically, a countercurrent on each side of the jet.

To carry the argument much further becomes difficult, because one has to allow for the possibility that changes in the stratification affect the mixing length in some way about which we can only guess. Since we do not know what these laws actually are, we may easily be led astray. Rossby (1936 a) is led to argue that there should be a countercurrent on the inshore side of the Gulf Stream, but none on the Sargasso Sea side. In fact, the countercurrent observed on the inshore side is very much weaker than the pronounced countercurrent on the Sargasso Sea side. Rossby's treatment of the dynamics of the countercurrents, in the wake-stream model, has been very stimulating. We shall use the same dynamical equations later in this chapter, when we come to discuss the Stream as a stream of uniform potential vorticity.

LATERAL MIXING

The Rossby wake-stream theory and the linear theories discussed in the preceding chapter all depend heavily upon the existence of large-scale turbulence: lateral mixing and the associated lateral-eddy viscosity. Whenever we attempt to frame a hydrodynamical problem in terms of eddy viscosity we must prescribe a certain scale of motion, to divide motions into what we will regard as *mean* motions and *turbulent* motions. Our dynamical equations are expressed explicitly in terms of the *mean* motions only; the *turbulent* motions are lumped statistically into one parameter: the lateral-eddy viscosity. It is clear that in a medium in which all scales of motion are present the distinction between mean motion and turbulence is purely arbitrary, but that it is of utmost importance to have a clear mental picture of the scale of motion which divides the two forms of motion in every theoretical discussion. Thus, as we have seen, the actual Gulf Stream meanders; a space and time average over several months would doubtless look rather like the gradual fanning out past Cape Hatteras that both Munk and Rossby seem to have in mind. Individual eddies and meanders would be statistically averaged out as turbulence; we may speak of the mean motion in this situation as the *climatological-mean Gulf Stream*. The pertinent scale of motion which separates mean motion from turbulent motion in this case is about 100 km., the half wavelength of meanders. As I have shown (Stommel, 1951, 1953), it seems likely that the eddies which break off from the Stream (for example, that represented in fig. 30) carry enough momentum to produce lateral shearing stresses of the order which Munk has postulated. But this is true only if we regard Munk's theory as applying to the climatological-mean Gulf Stream.*

Actually, of course, we are not satisfied with such a coarse statistical

*But see comment on pages 196-197 below.

description of the Stream as the one to which the 100 km. scale of averaging limits us. We want to describe more minute features of the Stream: a 5 or 10 km. scale is more appropriate to our purpose. In this case, of course, the meanders and eddies cannot be regarded as turbulence, and the mean motion exhibits the structure observed by a single hydrographic section: we speak of this narrow filament of moving water as the *instantaneous Gulf Stream*. It is the Stream described in Chapter V.

In attempting to frame a theory of the instantaneous Gulf Stream we do not have meanders and eddies at our disposal to provide large lateral stresses. As was mentioned in an earlier chapter (pp. 62 ff.), water-mass analysis indicates that there is not much mixing on a scale smaller than 10 km. In an attempt to obtain some quantitative information on lateral shearing stresses, I recently (Stommel, 1955 b) analyzed a set of current observations made by Pillsbury in the Florida Straits, and found that the coefficient of eddy viscosity was probably less than 10^6 cm.^{2}/sec., a value less than 2 per cent of the coefficient employed by Munk. Of course, these observations were not made in the Stream beyond the Straits, and the coefficient for that region is the one which we are really interested in knowing, but there are no long series of observations made there. Thus it seems difficult to discover enough turbulence of a scale smaller than 10 km. to provide important lateral-eddy stresses in the instantaneous Gulf Stream. We should emphasize here that apparently Munk was under the impression that he was dealing with the instantaneous Gulf Stream; his comparisons with hydrographic data indicate that. As I have suggested here, however, there is reason to suppose that his theory really applies to the climatological-mean Gulf Stream. We shall therefore attempt to frame a theory of the instantaneous Gulf Stream which does not take account of any important friction, but treats the Stream as essentially an inertial phenomenon.

Before leaving this vexed question of the importance of lateral stresses we should note an argument proposed by Rossby (1936 a) and Montgomery (1940) and recently questioned by Morgan (1956). Rossby asserted that since there is little bottom friction in the ocean, the torque of the wind stress applied to the surface of the sea can *only* be balanced by the torque of lateral shearing stresses around the coasts of the sea. Morgan (1956) has shown that there are other torques that need to be considered: a Coriolis torque, and a torque produced by differences of pressure at different parts of the coast. The relative importance of these various torques has not yet been ascertained; but much of the moral support for the indiscriminate use of lateral mixing and eddy coefficients has been removed.

Let us now explore some ideas which may provide an explanation of the detailed structure of the Gulf Stream independent of friction.

THE POTENTIAL VORTICITY OF A LAYER
OF UNIFORM DENSITY

In order to include nonlinear dynamical terms and density stratification, it is helpful to introduce a quantity called potential vorticity, obtained in the following manner. Using the coördinate systems employed in the preceding chapter, let us consider a homogeneous layer of fluid with density ρ_1 and thickness D, in which case the dynamical equations are

$$\frac{du}{dt} - fv = -\frac{1}{\rho_1}\frac{\partial p}{\partial x} + X,\tag{1}$$

$$\frac{dv}{dt} + fu = -\frac{1}{\rho_1}\frac{\partial p}{\partial y} + Y,\tag{2}$$

where $d/dt = (\partial/\partial t) + u(\partial/\partial x) + v(\partial/\partial y)$, and X and Y include both frictional driving and retarding forces, which we are not now interested in considering in detail. The Coriolis parameter f is regarded as a function of y. First, we eliminate the pressure gradients by cross-differentiation:

$$\frac{d}{dt}(f + \zeta) + (f + \zeta)\left(\frac{\partial u}{\partial x} + \frac{\partial v}{\partial y}\right) = \frac{\partial Y}{\partial x} - \frac{\partial X}{\partial y},\tag{3}$$

where ζ, the *relative vorticity*, is defined as $\zeta = (\partial v/\partial x) - (\partial u/\partial y)$. The quantity $f + \zeta$ is called the *absolute vorticity*. The equation of continuity of the layer is

$$\frac{dD}{dt} + D\left(\frac{\partial u}{\partial x} + \frac{\partial v}{\partial y}\right) = 0.\tag{4}$$

Elimination of the horizontal divergence between the two equations results in the equation of *potential vorticity*, $(f + \zeta)/D$,

$$\frac{d}{dt}\left(\frac{f + \zeta}{D}\right) = \frac{1}{D}\left(\frac{\partial Y}{\partial x} - \frac{\partial X}{\partial y}\right).\tag{5}$$

If there are no frictional forces, this equation simply states that the potential vorticity of a water column within the homogeneous layer cannot change as it moves from one place to another. Thus, as a column moves from one latitude to another, an adjustment must occur in the depth of the layer and in the relative vorticity in such a way as to keep the potential vorticity constant.

THE POTENTIAL VORTICITY IN AN
ISOPYCNAL LAYER

Any actual Gulf Stream cross section exhibits a marked decrease in the thickness of layers (bounded by isotherms) in the high-velocity parts of the current. This vertical shrinking is particularly noticeable in the isothermal layers warmer than 16° C. It seems dynamically significant that this vertical shrinking is just about the amount required for the potential vorticity of each isothermal layer to be uniform across the Stream, all the way from the inshore edge of the Stream, where vertical shrinking is at a maximum, to the Sargasso Sea. At the left-hand edge there is evidently a sharp discontinuity in potential vorticity. This is also true of the T-S correlations of the surface water.

In order to demonstrate this remarkable uniformity (Stommel, 1955 a) of potential vorticity, let us consider the vertical thickness h of the layer bounded by the 17 and 19° C. isotherms in Worthington's Gulf Stream section shown in fig. 33. We place the y-axis in the direction of flow of the stream, and direct the x-axis toward the right of the current, extending eastward into the Sargasso Sea.

In the Sargasso Sea, on the right-hand side of the section, the depth of the isothermal layer is constant, and we denote it by h_0. The velocity of the current vanishes there. Moreover, since the current is all in the y-direction, the relative vorticity is essentially given by $\partial v/\partial x$. The equation for uniform potential vorticity may then be expressed in the following way:

$$\frac{f+\frac{\partial v}{\partial x}}{h} = \frac{f}{h_0}. \tag{6}$$

The velocity v of the water in this isothermal layer may then be computed from the observed values of thickness h by direct integration:

$$v(x) = \int_x^\infty f\left(\frac{h}{h_0}-1\right) dx. \tag{7}$$

This computed velocity may now be compared with the geostrophically computed velocity. The results of such a comparison are shown in fig. 65. The fact that the velocities computed in these two very different and dynamically independent ways agree is good evidence that the potential vorticity in the Stream is nearly independent of cross-stream position.

MODEL WITH UNIFORM POTENTIAL VORTICITY

It is interesting to see how close a representation of the Gulf Stream can be obtained by using nothing but the principle of conservation of potential vorticity and the geostrophic approximation in a simple two-layer model. Taking the axes as before, we assume that a resting layer of uniform depth D_0, and density ρ_1, on top of another resting layer of very great depth, and density ρ_2, extends indefinitely in the positive x-direction. We now suppose that the interface between the two layers comes to the surface along the y-axis; thus $D=0$ at $x=0$. Therefore there must be a geostrophic current v

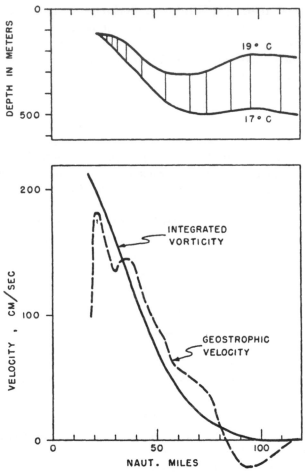

Fig. 65. Vorticity analysis of Worthington's section (1954b; see figs. 33 and 34 of the present study). The vertical thickness of the 17–19° layer is indicated in the upper graph. In the lower graph, the velocity computed by the vorticity formula (7) is shown by the solid curve; the geostrophic velocity as computed by Worthington, by the other curve. The fact that these two curves are fairly similar suggests that the potential vorticity in the upper layers of the Gulf Stream is fairly uniform across the Stream. For the dynamical significance of this, see the text.

(Gulf Stream) in the upper layer just to the right of $x=0$, but, by hypothesis, it must not extend very far into the central Atlantic (that is, $v=0$ for large x). Since the lower layer is regarded as being at rest everywhere, the current velocity is given by the geostrophic relationship in terms of D:

$$fv = g'\frac{\partial D}{\partial x}, \tag{8}$$

where $g' = g(\rho_2 - \rho_1)/\rho_2$. The configuration we have conjured up thus far does specify the total transport T of the stream,

$$T = \int_0^\infty vD\,dx = \frac{g'}{f}\frac{D_0^2}{2}, \tag{9}$$

but does not specify the shape of the stream profile, the width of the stream, or whether countercurrents, and so forth, exist. The stream might be wide or narrow. Therefore some further physical statement must be introduced, to determine the stream completely. Suppose that we now require that the potential vorticity of the upper layer be the same for all x:

$$\frac{f + \dfrac{\partial v}{\partial x}}{D} = \frac{f}{D_0}. \tag{10}$$

Combining this requirement with the geostrophic equation, we are led to the following equation in D:

$$\frac{\partial^2 D}{\partial x^2} = \frac{1}{\lambda^2}(D - D_0), \tag{11}$$

where $\lambda^2 = g'D_0/f^2$. Therefore the cross-stream profile of depth and velocity is completely determined:

$$D = D_0(1 - e^{-x/\lambda}),$$
$$v = \sqrt{(g'D_0)}\,e^{-x/\lambda}. \tag{12}$$

If we now introduce numerical values $D_0 = 800$ m., and $(\rho_2 - \rho_1)/\rho_2 = 2 \times 10^{-3}$, the model gives a realistic transport, $T = 64 \times 10^6$ m.³/sec. The maximum velocity of the Stream is 4 m./sec. The width of the Stream, taken as the distance between the point of maximum velocity and that at which the velocity is reduced to $1/e$ of the maximum, is 40 km. The length, λ, has been called by Rossby (1936a) the *radius of deformation*. Rossby introduced it in the presentation of his wake-stream theory, to explain countercurrents. Today it would seem that this term may actually be suitable for describing the Stream itself.

Since this very crude model of uniform potential vorticity is independent of y, that is, of latitude, there is no provision for the growth of the stream, such as is observed in the real Gulf Stream between the Florida Straits and Cape Hatteras. In the actual Gulf Stream the water present at any latitude y' is mostly drawn from lower latitudes, and hence one might reasonably ask why there should be any reason to expect the potential vorticity in the Stream at y' to correspond to that outside the Stream at y'. The justification for applying the simple model described rests on the observed fact that in the central Atlantic between 10 and 35° N. latitude, the potential vorticity of the top isothermal layers (> 10° C.) is remarkably constant. Suppose, for example, that the quantity D_0 is the thickness of the layer between the sea surface and the 10° C. isothermal surface. The depth of the 10° C. isotherm is shown in fig. 66. Since the relative vorticity in central oceanic regions is small, the reader may compute f/D_0 for the region between 10 and 35° N.

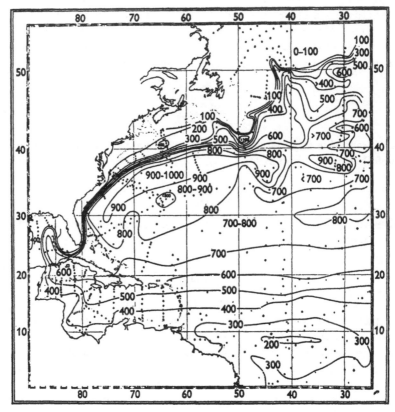

Fig. 66. Depth of the 10° C. isothermal surface in the western North Atlantic Ocean, according to Iselin (1936, fig. 47). Depths are given in meters.

and convince himself that over this large part of the ocean the potential vorticity so defined is actually uniform and nearly constant. From a theoretical point of view it might be preferable to use density structure (see figs. 10–15).

Dr George Morgan has pointed out to me that the easiest way to see the difference between the linear theories (the Munk theory, for example) and possible nonlinear theories is to interpret the potential-vorticity equation in the following way. The linear theory is steady, and essentially approximates the total derivative of the potential vorticity, $(d/dt)\,[(\zeta+f)/D]$, by $\beta v/D$. Thus ζ is regarded as small, and no account is taken of variations in D. The left-hand member of the potential-vorticity equation, formula (5) of this chapter, must balance against something, and hence all that remain are the various possible types of frictional terms.

In the high-velocity regions of the Gulf Stream, the relative vorticity ζ is negative and is not at all negligible as compared to f; inspection of the surface velocity profiles in fig. 32 will show the truth of this statement. Therefore it is quite conceivable that as a fluid column moves into the Gulf Stream, the changes in ζ and D can occur in such a way as to keep the potential vorticity constant and the conservation equation satisfied without the use of the frictional terms. Thus the theoretical model of a frictionless, nonlinear stream may describe the real Gulf Stream, the essential balance of terms within an isopycnal layer being of the following form:

$$u\left[\frac{(\zeta+f)}{D}\frac{\partial D}{\partial x}-\frac{\partial \zeta}{\partial x}\right]+v\left[\frac{(\zeta+f)}{D}\frac{\partial D}{\partial y}-\frac{\partial \zeta}{\partial y}-\beta\right]=0. \tag{13}$$

A very interesting series of free, steady solutions of the motion in a frictionless, homogeneous ocean of constant depth has recently been obtained by Fofonoff (1954). The conclusion drawn from his work is that eastward-flowing currents cannot be broad, slow streams, but must be narrow and have high velocity and high relative vorticity. As yet, Fofonoff has not been able to obtain similar solutions for a two-layer ocean in which horizontal divergence is appreciable, but he has indicated that, in the North Atlantic, this will produce a westward, as well as a northward, intensification. When these solutions have been obtained we should expect the profiles of the strong currents to be similar to those obtained earlier in this chapter, in the consideration of conservation of potential vorticity. Fofonoff's study is very similar to the uniform potential-vorticity model described above, and the models by Charney and Morgan described below. His paper (1954) was somewhat formal, however, and its applicability to the Gulf Stream becomes obvious only in the light of these and other studies.

In formulating the physical model for a nonlinear steady-state theory of ocean currents we shall try to preserve the following characteristics.

i. The ocean will be density-stratified.

ii. The solution for central ocean transport will be the same as Sverdrup's and Munk's.

iii. The narrow western stream (Gulf Stream) will be frictionless and convergent, and there will be a tendency toward development of a Fofonoff-type eastward flow in the region of the North Atlantic Drift.

iv. All dissipation of the energy in the Stream will occur in a limited region toward the end, where meanders break up and multiple streams form.

With respect to this region, the theory need not involve an assumption concerning lateral-eddy viscosity. The dissipation may be determined dynamically, as in a hydraulic jump in ordinary open-channel flow.

THE REGION OF DECAY OF THE GULF STREAM

After the Stream passes Cape Hatteras, it develops meanders, which increase in amplitude and eventually form detached eddies east of 60° W. longitude. The interpretation of observational data beyond this point is quite ambiguous, as has been mentioned in Chapter V. All indications are that the Stream breaks down into fragments and eddies which cannot be adequately described or surveyed by present means. Of all the regions of the Stream, this region of decay is the most difficult to depict and to understand. The region must be very extensive. The chart of the 10° C. isotherm depth (fig. 66) demonstrates that the potential vorticity of the upper layer is not even remotely uniform north of 35° N. latitude. The fragments, or, as they used to be called, 'branches', of the Gulf Stream still discernible in the North Atlantic Current (fig. 42) do not have the same potential vorticity, although there does appear to be a fair degree of uniformity within and to the south of any particular fragment. Apparently there are at least two complicated processes at work in this region which are not easily susceptible of simple theoretical treatment:

i. The Stream becomes unstable and breaks up into fragments. Each fragment seems to carry an equal share of the transport. The potential vorticity of each fragment is uniform, but different from that of the parent Gulf Stream. These fragments can be traced at least halfway across the North Atlantic. They underlie the surface West Wind Drift.

ii. Surface cooling and wind mixing profoundly modify the stratification of the surface layers, especially in the wintertime. Since, on the average, it takes more than a year for particles to pass entirely through the region of decay, it is reasonable to suppose that the potential vorticity of the waters in the region will be greatly changed by climatic factors.

In the course of his fundamental studies of the atmospheric jet stream, Rossby and his colleagues at the University of Chicago have developed many interesting ideas which may have some bearing on the multiple-stream structures in the region of decay (1947). Neumann (1952) has also studied the region, and believes that the multiple streams are not decaying fragments of the Stream; he attributes them to persistent nonuniformities of the winds. My own impression is that wind patterns of the sort required do not persist long enough to produce such effects.

THE CRITICAL INTERNAL FROUDE NUMBER

One of the important dimensionless numbers characterizing the flow of homogeneous frictionless fluids in channels with a free surface is the well-known Froude number, U^2/gD, where U is the mean velocity of the fluid along the channel, g is the acceleration due to gravity, and D is the distance from the free surface to the bottom of the channel. According to ordinary hydraulic engineering practice, the flow is said to be subcritical, critical, or supercritical, depending upon whether the Froude number is less than, equal to, or greater than unity. In natural watercourses and engineering works such as canals and aqueducts, the flow is almost always subcritical except at certain control points (weirs and dams, for example), where the flow is locally critical. Whereas small perturbations in water level (long gravity waves) can move upstream in subcritical flow, they are brought to a standstill at points along the watercourse where the flow is critical. This explains the importance of points of critical flow in determining the water level upstream. Supercritical flow in nature is comparatively rare; it normally terminates abruptly in a hydraulic jump which converts the flow back to subcritical.

Considering the great depth of the ocean, and the very moderate velocities of flow, it is clear that no ocean current even remotely approaches the ordinary critical Froude number. However, in a density-stratified fluid there are other 'internal' Froude numbers associated with the speed of propagation of long internal waves. In a system consisting of a very deep dense layer at rest, and a less dense surface layer of depth D moving at velocity U, the internal Froude number is $U^2/g'D$, where $g' = (\Delta\rho/\rho)g$. In the stream of uniform potential vorticity discussed earlier in this chapter, in the theories of Morgan and Charney outlined later in this chapter, and in the simple theory of steady meanders presented in Chapter IX, parts of the flow are supercritical with respect to the internal Froude number. In the summer of 1955, in discussions at the Woods Hole Oceanographic Institution, Dr Rossby pointed out that such regimes are likely to be unstable and probably do not exist in nature. Therefore we should not expect transverse profiles of velocity such as those given in equation (12) to hold for values of D/D_0 less than that for which $v = \sqrt{(g'D)}$. An easy calculation, for the stream of uniform vorticity, shows that this occurs at $D = 0.38 D_0$. In view of the fact that this simple model of the Gulf Stream fits the data from observations so well, it is interesting to note that this is where the point of inflection in the depth of the $10°$ C. isotherm actually occurs (see fig. 33). Rossby (1951) has extended the concept of critical flow to flows with arbitrary vertical density structure.

Up to the present, no one has succeeded in working out a detailed Froude theory which also takes account of the earth's rotation. However, it does seem likely that the effect of the earth's rotation will not radically alter the internal Froude criterion: the real Gulf Stream approaches critical internal Froude flow, and it is quite conceivable that as a result there are internal hydraulic jumps and other interesting small-scale phenomena, such as oblique shock fronts, along the left-hand, inshore, edge of the Stream.

THE BOUNDARY-LAYER TECHNIQUE

Although Munk's first theoretical treatment of the wind-driven ocean currents involved complete analytical solutions for the entire ocean, he quickly realized that, since the viscous terms (the highest-order derivatives) were important only near the coasts, the problem could have been solved more simply by using Sverdrup's (1947) simple first-order (in transport function) equations for the central parts of the ocean, and fitting a viscous boundary layer to the solution for the central regions at those places near the coasts where higher-order terms become significant. In the more complicated cases of triangular ocean basins this alternative procedure was adopted (Munk and Carrier, 1950). The ocean is thus divided into two regions: one, which we shall call the *interior* region, which includes all central regions of the ocean, and which is determined by the form of the wind stress; the other, the *boundary-layer* region, in which the higher-order terms are important. Actually, so far as the boundary layer is concerned, it does not matter what physical process produces the *interior* current regime. As long as some interior solution exists, be it wind-driven or thermodynamically produced, a linear viscous boundary layer can be fitted to it on the western coast.

We shall take advantage of this natural separation of the problem into two parts in our formulation of the Gulf Stream as an inertial boundary layer. Furthermore, we shall limit the application of the theory to the region bounded by latitudes 10 and 35° N., the region of formation or growth of the Gulf Stream. North of 35° N. latitude the Stream breaks away from the coast, and we enter the region of decay. This region is not dealt with in either the Morgan theory or the Charney theory, both of which are discussed in the next section.

THE INERTIAL THEORIES OF MORGAN
AND CHARNEY

In the latter half of 1954 I made several futile attempts to construct a theory of an inertially governed Gulf Stream that would provide a more complete solution than the partial solutions described above. I confided my troubles

to Dr Jule G. Charney of the Institute for Advanced Study at Princeton, and
to Dr George W. Morgan of Brown University, and in the spring of 1955
these two investigators, independently, developed proper inertial theories
of the Stream. Their theories lead to a nonlinear differential equation which
can be integrated numerically.

Let us first examine Morgan's formulation. We consider an ocean bounded
by meridional walls at $x=0$ and $x=r$, as in Chapter VII. For simplicity we
also assume that the wind field over an interval of latitude $0 \leqslant y \leqslant s$ is given
by the equation

$$\tau_x = -\tau_0 \left(1 - \frac{y^2}{s^2}\right), \quad \tau_y = 0. \tag{14}$$

According to the dynamical equation applying to the central part of the
ocean [Chapter VII, equation (42); also Chapter XI, equation (5)] we can
solve for the meridional transport per unit width, M_y, in the interior:

$$M_y = \frac{\partial \psi}{\partial x} = -\frac{2y\tau_0}{\rho \beta s^2}. \tag{15}$$

On the east coast, $x=r$, the zonal transport $M_x = -\partial \psi/\partial y$ vanishes;
hence we have, from continuity, the following simple results for ψ and M_x:

$$\psi = \frac{2y\tau_0}{\rho \beta s^2}(r-x), \tag{16}$$

$$M_x = -\frac{2\tau_0}{\rho \beta s^2}(r-x). \tag{17}$$

The advantage of the parabolic zonal wind-stress law, equation (14), is
that it specifies a zonal transport at the western coast, $x=0$, independent of
the latitude y. Solutions (15), (16), and (17) are independent of any assumption about the vertical density stratification. If we consider a homogeneous
ocean, the variation in total depth due to slope of the free surface is negligible. If, on the other hand, we consider a two-layer system with density
different in each layer, and the bottom layer at rest, we must also take into
account the variation of the depth, D, of the top layer. In particular, the
pertinent dynamical equation in the interior is the y-equation:

$$-fM_x = -g' \frac{\partial}{\partial y}\left(\frac{D^2}{2}\right). \tag{18}$$

Since, for the purpose of matching the boundary layer to the interior
solution, we need only the value of D along the outer edge of the boundary
layer (that is, for $x=0$), we can integrate equation (18) with respect to y,
using M_x at $x=0$ from equation (17) and taking into account the variation
with latitude of the Coriolis parameter with latitude. We set $f=f_K+\beta_K y$.

In this way we obtain the following expression for use in the two-layer system:

$$D^2(0, y) = \frac{2M_x}{g'}\left(f_K y + \frac{1}{2}\beta_K y^2\right) + D^2(0, 0).$$ (19)

The quantity $D(0, 0)$, the interior depth of the interface at the outer edge of the boundary layer, and at $y=0$, is a constant which we can choose arbitrarily so far as this theory is concerned. In this way we can form a variety of models.

We now turn to the boundary layer itself. Morgan (1956) uses the following dynamical equations:

$$fv^* = g'\frac{\partial D^*}{\partial x},$$ (20)

$$\frac{v^{*2}}{2} + g'D^* = B(\psi^*).$$ (21)

The asterisks are used to denote quantities defined within the boundary layer. Equation (20) states that the dynamical equation for force components normal to the coast is essentially geostrophic. Equation (21) is Bernoulli's (nonlinear) equation for the upper layer. Next to the coast, u^{*2} is insignificant compared to v^{*2}. In the boundary layer the kinetic plus potential energies are a function $B(\psi^*)$ of the transport function only. Outside the boundary layer this is not so, because the wind stress has a long time to do work on the water there. It is convenient to transform these two equations into a form containing only ψ^* and D^*, by making use of the definition $v^*D^* = \partial \psi^*/\partial x$:

$$\psi^* = \frac{g'}{2f}D^{*2} + C(y),$$ (22)

$$\frac{\partial \psi^*}{\partial x} = \sqrt{2}\, D^*\, [B(\psi^*) - g'D^*]^{1/2}.$$ (23)

The quantity $C(y)$ is easily evaluated along the outer edge of the boundary layer by matching ψ^* and D^* with interior solutions ψ and D at $x=0$ [equations (16) and (19)]. In this way Morgan obtains the result:

$$C(y) = \frac{\beta_K y^2 M_x(0, 0)}{2f} - \frac{g'D^2(0, 0)}{2f},$$ (24)

where we write $M_x(0, 0) = M_x(0, y)$. When he substitutes this in equation (22) and solves for D^{*2}, Morgan obtains

$$D^{*2} = D^2(0, 0) + \frac{2f}{g'}\psi^* - \frac{\beta_K y^2 M_x(0, 0)}{g'}.$$ (25)

Another relation between ψ^* and D^* can be obtained from the Bernoulli equation (21), because at the outer edge of the boundary layer $v^{*2} \to 0$, and hence

$$g'D^* = B(\psi^*). \tag{26}$$

Thus the function $B(\psi^*)$ is determined by

$$B[\psi(0, y)] = g'[D(0, y)] = g'\left[D^2(0, 0) + \frac{2f_K\psi^*}{g'} + \frac{\beta_K}{g'M_x(0, 0)}\,\psi^{*2}\right]^{1/2}. \tag{27}$$

When this value of $B(\psi)$ is substituted in equation (21) and D^* is eliminated by equation (25), the following equation is obtained:

$$\left(\frac{\partial\psi^*}{\partial x}\right)^2 = 2g'\left[D^2(0, 0) + \frac{2f_K}{g'}\,\psi^* - \frac{\beta_K y^2 M_x(0, 0)}{g'}\right]$$

$$\times\left\{\left[D^2(0, 0) + \frac{2f_K}{g'}\,\psi^* + \frac{\beta_K}{g'M_x(0, 0)}\,\psi^{*2}\right]^{1/2}\right. \tag{28}$$

$$\left. - \left[D^2(0, 0) + \frac{2f_K}{g'}\,\psi^* - \frac{\beta_K y^2 M_x(0, 0)}{g'}\right]^{1/2}\right\}.$$

The proper boundary condition is that $x = 0$ be the streamline $\psi^*(0, y) = 0$. It is important to note that if $\beta_K = 0$, the quantity $\partial\psi^*/\partial x = 0$ for all x. Thus, no boundary layer forms if the variation of the Coriolis parameter with latitude is neglected. Morgan proceeds to analyze the nonlinear differential equation (28) by introducing dimensionless variables, $\bar{x}, \bar{y}, \overline{\psi}^*$:

$$x = s\bar{x}, \quad y = s\bar{y}, \quad \psi^* = M_x(0, 0)\,s\overline{\psi}^*; \tag{29}$$

and the parameters ϵ and δ are defined as follows:

$$\epsilon = \frac{D^2(0, 0)\,g'}{M_x(0, 0)\,\beta_K s^2}, \tag{30}$$

$$\delta = \frac{2f_K}{\beta_K s}. \tag{31}$$

The differential equation (28) thus may be written

$$\left(\frac{\partial\overline{\psi}^*}{\partial\bar{x}^2}\right)^2 = \frac{2\beta_K^{3/2}s^3}{M_x^{1/2}(0, 0)\,g'^{1/2}}[\epsilon + \delta\overline{\psi}^* + 2\bar{y}\overline{\psi}^* - \bar{y}^2]$$

$$\times\{[\epsilon + \delta\overline{\psi}^* + \overline{\psi}^{*2}]^{1/2} - [\epsilon + \delta\overline{\psi}^* + 2\bar{y}\overline{\psi}^* - \bar{y}^2]^{1/2}\}. \tag{32}$$

For numerical integration, the involved numerical coefficient in equation (32) can be absorbed into the independent variable by the transformation

$$\xi = \left(\frac{\beta_K^{3/2}s^3}{M_x^{1/2}(0, 0)\,g'^{1/2}}\right)^{1/2}\bar{x}. \tag{33}$$

Morgan (1956) gives his numerical results in several graphs, from which figs. 67 and 68 have been redrawn. Fig. 67 shows the variation of the dimensionless transport function $\overline{\psi}^*$ with the dimensionless distance $\overline{x}$ from the coast $\overline{x} = 0$ at $\overline{y} = 1$. The value chosen for β_E in all these calculations was 2×10^{-13}/cm. sec. The magnitudes chosen for other parameters are given in the legend to fig. 67. Except for the fact that the boundary of the southern

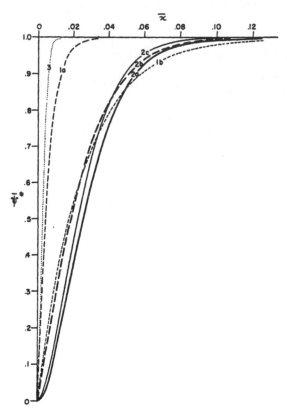

Fig. 67. Dimensionless plot of mass-transport function for several different cases, according to Morgan (1956).

Type of ocean model	Curve	$M_x(0, 0)$ cm.³/sec.	$D(0, 0)$ cm.	s cm.	g' cm./ sec./ sec.	ϵ	δ
Homogeneous ..	1a	10^5	4×10^5	2×10^8	10^3	2×10^5	0
	1b	10^5	2×10^4	2×10^8	10^3	5×10^2	0
Two-layer......	2a	10^5	2×10^4	2×10^8	2	1	0
	2b	10^5	2×10^4	2×10^8	4	2	0
	2c	10^5	2×10^4	2×10^8	2	1	1
	3	5×10^4	1.2×10^5	6×10^8	0.25	1	0

limit of the interior region of the model indicated by curve 2a is at the equator ($\delta = 0$), curve 2a corresponds to a model very much like the real Gulf Stream. Morgan finds the width of the Stream to be 150 km. Curve 2c represents the same situation with $\delta = 1$, that is, with the southern boundary at approximately 15° N. latitude.

The shift of the southern boundary produces a slightly narrower stream, but the effect is quite small. The parameters chosen for curve 1b represent a homogeneous ocean with a depth equal to that of the upper layer of model 2a. One sees that the stratification does not have much influence on the shape of the stream. Comparison may also be made between curve 2a and curve 1a. Curve 1a represents a homogeneous ocean the depth of which is like that of the real ocean; this is the stream which would exist if the ocean were homogeneous. It is even narrower than the real Stream.

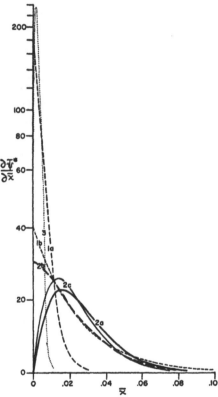

Fig. 68. Dimensionless plot of the transport per unit width as a function of distance from the western coast, according to Morgan (1956). The numbers on the curves refer to the same choice of parameters as those shown in fig. 67.

A further indication of the insensibility of the Stream to density stratification shows up by comparing curves 2a and 2b. The density difference in the latter is twice that in the former; otherwise, all parameters are the same.

Curve 3 shows a very narrow stream, the narrowness being due to the increase in s. It applies to inertial boundary layers that might be formed in connection with the pole-to-pole thermohaline circulation discussed in Chapter XI.

In fig. 68 the dimensionless transport per unit width, $\partial \bar{\psi}^*/\partial \bar{x}$, is plotted against $\bar{x}$ for each of the cases discussed above. The transport tends to decrease monotonically with increasing x because of the decrease in velocity, but this is partly offset by the small D^* in the two-layer models very close to $x = 0$; thus in these cases the transport first increases and then decreases with increasing $\bar{x}$. In this way the streams shown by curves 1b and 2a are of about the same width, but the latter is displaced somewhat away from the coast.

Although Morgan's curves indicate that stratification does not change the form of the stream in an inertial boundary layer, there is an important point that, as he has pointed out, needs emphasis. First, let us write equation (25) in terms of the dimensionless variables:

$$D^{*2} = D^2(0, 0) + \frac{2(f_K + \beta_K s\bar{y})}{g'} M_x(0, 0)\,\overline{\psi}^* - \frac{M_x(0, 0)\beta_K s^2\bar{y}^2}{g'}. \qquad (34)$$

Thus $M_x(0, 0)\beta_K s^2/g'$ is a measure of the change of depth along a streamline, and ϵ is the ratio of the characteristic depth $D^*(0, 0)$ to the characteristic change in depth. For fixed $\bar{y}$ the depth is smallest at the coast. It vanishes when

$$\bar{y} = \epsilon \quad \text{or} \quad y = \frac{D^{2*}(0, 0)g'}{M_x(0, 0)\beta_K s}. \qquad (35)$$

Thus Morgan shows that if $\epsilon < 1$, the solution cannot be valid for $\bar{y} > \epsilon$, and the value $\bar{y} = \epsilon$ might be regarded as a latitude north of which a new regime of flow must occur. The region of decay, of meanders and eddies beyond Cape Hatteras, comes to mind.

There are two further remarks that we can make concerning what, in nature, this limiting condition (35) may imply. First, it should be remembered that the warm water of the upper layer is produced in the interior by climatic influences acting upon the sea surface in subtropical and equatorial regions. Therefore it seems reasonable to suppose that the interface (or thermocline) ought to come to the surface in a latitude between the latitude of maximum curl of the wind stress and that of the northern limit of the subtropical gyre—between 35 and 50° N. latitude. In other words, the choice of $D(0, 0)$ is arbitrary only so long as we do not include climatic influence in our models. Furthermore, if $D(x, y) = 0$ does not coincide fairly well with $\psi(x, y) = 0$ in the region of maximum westerly wind stress, water of the upper layer is not conserved in the upper layer. Of course, the real ocean is not a simple two-layer system, and we cannot demand complete conservation in reality, but the climatic influences acting upon the sea do not appear to be powerful enough to offset an excessive loss of surface water. Much of it must recirculate without major density modification.

Secondly, the quantity $D(0, 0)$ may be determined, to some extent, by the control action (in the hydraulic engineering sense) of the Gulf Stream itself. As the upper layer is made thinner and thinner, the velocities within it become higher and higher, until finally a minimum depth of that layer is reached beyond which the transport required by the interior wind-stress solution cannot be sustained [equation (35)]. It is at this very stage that the crude meander theory outlined in Chapter IX indicates the onset of meander instability. One is tempted to speculate, therefore, that the Gulf Stream automatically reaches the condition given by equation (35) at the latitude of maximum wind curl, and that it acts in much the same way as an ordinary open-channel transition (weir, overfall, and so on) to control the over-all depth of the thermocline in the North Atlantic Ocean.

One of the most gratifying features of the inertial theory of the Gulf Stream is that it does not contain an arbitrary parameter such as the viscous boundary-layer theory does. A disconcerting feature is the absence of a well-defined countercurrent to the east of the Stream. Formally this is accounted for by the lower-order derivatives in the inertial boundary-layer theory. Too much emphasis should not be placed upon this apparent advantage of the viscous over the inertial theory, because there is a bit of confusion among various investigators over the countercurrent to the east of the Stream: there are two 'countercurrents' of widely different nature. The first kind is apparently associated with the warm core. It is evidently a dynamical necessity resulting from advection of warm water from lower latitudes (see fig. 33). The simple two-layer model treats the upper layer as uniform, so of course it cannot show this feature. Since this countercurrent has a width similar to that of the main Gulf Stream, I assume it is the countercurrent which the Munk theory attributes to lateral friction.

The countercurrent of the second kind is a much deeper, broader feature of the circulation. It shows up somewhat in fig. 11, as indicated by the close spacing of the 800 and 1000 m. contours of the $27 \cdot 0$ σ_t surface along the $30°$ N. latitude circle. It is much too wide to be part of the boundary layer, and there is no feature of the wind-stress distribution which can account for it in the interior. Perhaps this countercurrent of the second kind reveals some fundamental discrepancy between theory and reality; just what it is, I do not know.

Whereas Morgan's treatment (1956) shows quite clearly the role of the various physical parameters in the determination of the Stream profile, it is not aimed at giving a detailed representation of the actual Gulf Stream. Charney's (1955) study constructs a two-layer ocean model with a choice of D and ψ on the outer edge of the boundary layer as near to what actually occurs in nature as is possible. The two studies thus supplement each other.

Charney's formulation makes use of the Bernoulli equation (21) and the equation of potential-vorticity conservation, equation (5). If we make use of the fact that in the boundary layer, $|\partial v^*/\partial x| \gg |\partial u^*/\partial y|$, and use the definition $v^* D^* = \partial \psi^*/\partial x$, the potential-vorticity equation is of the following form:

$$\frac{\dfrac{\partial}{\partial x}\left(\dfrac{1}{D^*}\dfrac{\partial \psi^*}{\partial x}\right) + f(y)}{D^*} = F(\psi^*).$$ (36)

The arbitrary function $F(\psi^*)$ is determined by matching functions at the outer edge of the boundary layer with observed data. Here we make use of the fact that the higher-order terms in equations (21) and (36) vanish at the outer edge of the boundary layer.

$$\frac{f}{D} = F(\psi^*), \tag{37}$$

$$g'D = B(\psi^*). \tag{38}$$

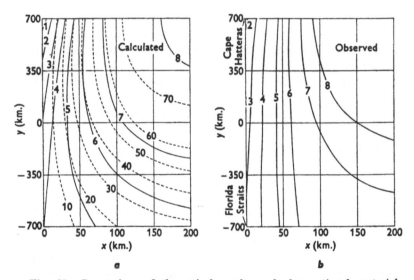

Fig. 69. Comparison of theoretical results and observational material, according to Charney's (1955) analysis. a, calculated depth D^* and mass-transport function ψ^* of the upper layer; contours of D^* (solid lines) are drawn for each 100 m., and streamlines (broken lines) for each 10×10^6 m.^{3}/sec. b, schematic chart of the observed mean depths of the 10° C. isotherm. The scale of the abscissa is exaggerated so that details of the field will show up better.

At the outer edge of the boundary layer D is prescribed by observation as a function of y. Since at this place

$$u = -\frac{1}{D}\frac{\partial \psi}{\partial y} = -\frac{g'}{f}\frac{\partial D}{\partial y},$$

we also obtain ψ as a function of y. Therefore, at the outer edge f/D and $g'D$ are known functions of y, and hence of ψ. In this way Charney (1955) determines $F(\psi)$ and $B(\psi)$ along the outer edge of the boundary layer, and therefore at every point inside the boundary layer which is connected with the outer boundary by a transport line.

In order to make a close comparison between the results of theory and those of observation, Charney has chosen a parabolic form for the transport function along the outer edge of the boundary layer, and has introduced an arbitrary Florida Straits transport. The depth of the interface in the two-layer model is identified with the depth of the 10° C. isothermal surface. The results of Charney's attempt to reproduce the flow pattern in the growth region of the Gulf Stream, between the Florida Straits and Cape Hatteras, are shown in fig. 69. Fig. 69, a, depicts the theoretically deduced contours of the thickness of the upper layer, D^*, in intervals of 100 m. (solid lines); and also contours of the transport function, ψ^*, at intervals of 10×10^6 m.^{3}/sec. (broken lines). The x-scale has been exaggerated by a factor of five. Fig. 69, b, is a schematization of the depths of the 10° C. isotherm taken from fig. 66. The agreement between the two charts is good, and, as Charney says, could doubtless be made even more complete by a numerical treatment involving actual coastal geometry, and real continuous density stratification along the outer edges of the boundary layer. At present such refinements do not seem worth while, however. One cannot help believing that a correct zero-order approximation has now been achieved.

Chapter Nine

MEANDERS IN THE STREAM

Very little is really known about the meanders of the Gulf Stream. They have been observed only a few times; beyond their wavelength, the direction of propagation, and the fact that they may grow into detached eddies, nothing is known. There are no detailed data on their structure as a function of depth, especially below the 900 ft. level commonly reached by the bathythermograph. However, because they are essentially wavelike they are tempting theoretically, and attempts have been made to treat them by linearized perturbation equations.

THE APPLICATION OF THE PERTURBATION METHOD

The perturbation method has been eminently successful in the classical theories[1] of sound waves, of tides, of gravitational water waves, and of the Bénard-type thermal-convection cells. Most of these problems are peculiarly suited to the perturbation method because they involve very simple models: the fluid is at rest, or in uniform motion, and the coefficients of the perturbation equations are constants. Helmholtz (1868) first investigated the following unstable system. Two homogeneous currents move, one above the other, with uniform (but different) velocities, under the influence of gravity. The upper layer is less dense than the lower. The shear at the common boundary tends to produce instability; gravity tends to produce stability. Short waves on the common boundary are shown to be unstable; long waves are stable.

Kelvin (1871) applied the perturbation technique to the problem of wind

---

[1] This historical section is after Queney (1950).

blowing over water and introduced capillarity as an additional stabilizing influence, which acts mostly on very short waves. As a result, the system is stable for all wavelengths of perturbation if the wind is below a certain critical speed. With increasing wind speed a single wavelength becomes unstable. Kelvin proposed to apply this theory to the formation of wind waves, but observation shows that real wind waves are more complicated than those envisaged in this simple theory. Even this problem, which at first seems so easy and for which there is no lack of observational material, has never been successfully resolved.

Meteorologists must deal with problems of a vastly more complicated nature when they attempt to apply the perturbation method to such phenomena as the development of waves and cyclones in the atmosphere. After J. Bjerknes and his co-workers discovered atmospheric fronts in 1919, they came to regard cyclones as unstable waves of the common boundary (inclined frontal surface) between two air masses with uniform (but different) densities and velocities. Tentative approximate solutions were given by V. Bjerknes, J. Bjerknes, Solberg, and Bergeron (1933). Owing to the fact that there are two stabilizing factors—gravity, operating now on short waves, and the earth's rotation, acting on the longest waves—the system is dynamically stable so long as the wind shear is less than a critical value, but when this value is just exceeded, a dominant unstable wave appears. This was interpreted as a cyclone.

The chief difficulty of this model is that it was soon discovered that fronts never exist outside of cyclones, a fact which made it seem unlikely that they could be the origin of cyclones. Some cyclones have no fronts associated with them at all. With the increase of meteorological data it has become increasing clear that cyclones are not the result of a degenerating mean zonal motion; indeed, Starr (1953) suggests that they are the cause of the mean motion.

Modern meteorologists have abandoned the notion that large-scale atmospheric perturbations are necessarily associated with discontinuities (fronts), and have turned to other models, such as that of Rossby's (1947) theory of long waves in the westerlies. Other present-day models involve vertical wind shears, and horizontal temperature gradients (see Fjørtoft, 1951, for a summary) of a form so remote from the structure of the Gulf Stream that they are not likely to be of help in understanding meanders.

When we come to consider the Gulf Stream, we are not embarrassed by the difficulty involved in early cyclone theory: there is no doubt that the Gulf Stream is the primary phenomenon, and that the meanders are secondary. Hence the Gulf Stream may surely be regarded as the basic undisturbed current, on top of which the meanders appear as perturbations.

On the other hand, we have so little consecutive data on the development of a meander into an eddy that we are at a distinct disadvantage. For example: after an eddy breaks off, does the parent stream mend itself, or do the two ruptured ends remain broken, the severed fragment drifting away to the northeast, and the fresh end penetrating to the east to form a new extension of the Gulf Stream?

GULF STREAM MEANDERS

The Gulf Stream meanders observed on the Multiple Ship Survey of June, 1950 (Fuglister and Worthington, 1951), have inspired several theoretical studies. It has been suggested that they are analogous to the waves in the atmospheric jet stream, but both their small scale (100–400 km.) and their small velocities of propagation suggest that the variation of the Coriolis parameter with latitude is not a dominant factor in their dynamics. Haurwitz and Panofsky (1950) have constructed several models of currents in a homogeneous ocean with cross-stream velocity profiles similar to those observed in the true Gulf Stream, and have carried out an intricate perturbation analysis to show the existence of unstable waves with reasonable velocities of propagation. These waves are a result of the shearing instability of the Stream.

In a series of lectures at Woods Hole in 1954 on the stability of large-scale motions Dr Jule Charney showed how application of some meteorological studies might be made to Gulf Stream meanders, in particular, the work of Kuo and Phillips. Kuo (1949) has made an extensive study of the instability of single smooth-profiled streams in a barotropic atmosphere (one in which the pressure and density surfaces are parallel). The wavelength of maximum instability in Kuo's theory is about 2·7 times the width of the stream, or about 140 km. Phillips (1951) has discussed the instability of a particular baroclinic model (pressure and density surfaces intersecting) consisting of two layers of equal thickness, one flowing over the other. For values of the parameters similar to those of the Gulf Stream, the maximum instability occurs for waves about 300 km. long, the rate of increase of amplitude, about three times in two weeks. Neither of these models resembles the real Gulf Stream very closely. Perhaps all we learn from them is that the Gulf Stream *may be* unstable, either barotropically or baroclinically, or both.

It seemed to me (Stommel, 1953) that certain types of meanders might exist in which stratification and inertia are dynamically important. In order to illustrate such a system, meanders in a very wide current were studied. To simplify the analysis, the realistic cross-stream velocity profiles of Haurwitz and Panofsky (1950) were abandoned. The density stratification of the real ocean is approximated by a two-layer system.

A SIMPLE MEANDER THEORY FOR A VERY WIDE CURRENT IN A STRATIFIED OCEAN*

Let us suppose that the lower layer is very deep, and hence that the horizontal pressure gradients vanish in it at all times. In the undisturbed state a steady current U flows in the x-direction in the upper layer of thickness D. Associated with this current is a cross-stream pressure gradient of the following form:

$$fU = -g' \frac{\partial D}{\partial y}. \tag{1}$$

We now suppose that there are small perturbations u, v, in the velocity components, and h, in the elevation of the free surface, and that these quantities are independent of y, the cross-stream coördinate. The perturbation equations may be written in the form

$$\left(\frac{\partial}{\partial t} + U \frac{\partial}{\partial x}\right) u - fv = -g \frac{\partial h}{\partial x}, \tag{2}$$

$$\left(\frac{\partial}{\partial t} + U \frac{\partial}{\partial x}\right) v + fu = 0, \tag{3}$$

$$\left(\frac{\partial}{\partial t} + U \frac{\partial}{\partial x}\right) \left(\frac{\rho}{\Delta\rho} + 1\right) h + D \frac{\partial u}{\partial x} + v \frac{\partial D}{\partial y} = 0. \tag{4}$$

If the perturbations are all in the form $e^{i(kx - \nu t)}$ we obtain the frequency equation

$$k^2 U^2 (1-p)^3 - \left[f^2 + gk^2 \frac{\Delta\rho}{\rho} D\right](1-p) + f^2 = 0, \tag{5}$$

where $p = c/U = \nu/kU = p' + ip''$, the real part p' of which is the ratio of the velocity of propagation of the wave to the velocity of the current, and the imaginary part p'' of which gives the instability of the wave motion. For the particular range of parameters involved, no one of these terms is small compared to the others. It is convenient to rewrite this equation in the form

$$y^3 + 2 = Py, \tag{6}$$

where

$$P = 2 \left(\frac{f^2}{2k^2 U^2}\right)^{1/3} \left(\frac{f^2 + g'k^2 D}{f^2}\right) \tag{7}$$

and

$$y = \left(\frac{f^2}{2k^2 U^2}\right)^{-1/3} (1 - p). \tag{8}$$

The roots of this equation are all real, provided $P > 3$, in which case there are three types of stable wave present. If $P < 3$ there is a region of unstable

*See also page 197 below.

waves. These are of most interest to us because they are the only ones likely to grow large enough to be noticed on a ship survey. Examination of the coefficients of the frequency equation reveals that for $U^2 < g'D$ all waves are stable. At $U^2 = g'D$ a single wave number given by $k = f/\sqrt{2}\,U$ becomes 'just unstable', whereas all other wave numbers are stable. For slightly larger values of U^2 there is a narrow range of wave numbers about $k = f/\sqrt{2}\,U$ in which waves are unstable. In the critical case of marginal stability the 'just unstable' wave is stationary.

One objection to applying this model to the meanders observed in the Gulf Stream is that the real Gulf Stream is not very wide. A more sophisticated theory would include lateral boundaries to the Stream and would provide for resting layers of water on each flank beyond the boundaries. Also, a perturbation theory applies only to waves of infinitesimal amplitude, whereas meanders often grow to large amplitude. Therefore it is important to regard this treatment of meanders as merely indicative of a possible mechanism for describing the meandering of a stratified current; but it is hard to see how it could be tested by actual observation of the crests and troughs of meanders, because the theory and observational techniques are both too crude to admit of a meaningful comparison.

Application to the Gulf Stream.—Rossby (1951) has shown that the velocity of the Gulf Stream does in fact approach the critical value $\sqrt{(g'D)}$. This is also true for the stream of uniform potential vorticity (see Chapter VIII). We might expect the Stream to become progressively shallower downstream as a result of friction, and gradually to approach the critical condition. Because of the paucity of hydrographic sections of the Stream in any one year or season, it is necessary to construct a composite series of sections, in order to determine whether there is any noticeable change in the depth of the Stream along its axis. A number of sections all made in early June of several years have been assembled and recomputed: (i) one at Hatteras at about 74° W.; (ii) two more on the Montauk Point–Bermuda line surveyed in Iselin's (1940) studies; (iii) one at 58° W.; and (iv) two Ice Patrol sections along the 50° W. meridian. The geostrophic transports at different depths were computed. In order to exhibit any changes in the depth of the Stream, the percentage of the total transport below certain selected depths was computed. The results are given in table 6; it is clear that there is no striking change in the depth of the Stream indicated by these sections. The inference to be drawn from this is that from Cape Hatteras to the tail of the Grand Banks the Stream is near to the critical velocity almost everywhere, and that unstable meanders might be expected anywhere.

We may ask ourselves what the size of the meanders predicted by our two-layer meander theory might be expected to be. A surface layer 200 m.

TABLE 6

PERCENTAGE OF TOTAL TRANSPORT BENEATH SELECTED DEPTHS AT
DIFFERENT JUNE HYDROGRAPHIC SECTIONS ALONG THE GULF STREAM
(2000 METER REFERENCE LEVEL)

Depth in meters	74° W. Off Cape Hatteras	68° W. Montauk Pt.– Bermuda line		58° W. Section	50° W. Sections	
	Stations					
	Atlantis	*Atlantis*		*Atlantis*	Ice Patrol	
	4565–4570	2867–2871	3054–3058	2616–2620	2715–2719	4175–4184
0	100	100	100	100	100	100
50	94	92	92	92	91	92
100	88	85	84	84	82	84
200	74	71	69	69	65	69
300	61	58	55	56	50	56
400	49	46	43	44	37	45
600	30	38	23	26	19	28
900	12	13	7	9	6	12
1200	4	7	2	3	2	4
1600	1	1	0	1	0	1
2000	0	0	0	0	0	0

Source: Stommel (1953, table 2).

thick, moving at 200 cm./sec. with a density difference of $\Delta\rho/\rho = 2 \times 10^{-3}$, is critical. The wavelength of the 'just unstable' perturbation corresponding to this choice of parameters is 180 km. All other wavelengths are stable and do not grow. It is interesting that this wavelength corresponds very closely to that of the large stationary meander observed to grow and detach into an eddy (see Chapter V), but other admissible choices of parameters would have led to poorer agreement.

It is important to emphasize that the meander theory presented here is not complete or proven, but merely suggestive of a type of wave motion which may possibly dominate the dynamics of meanders. If the energy of meanders is absorbed from the potential energy of the cross-stream mass distribution rather than from the kinetic energy of the flow, the meanders are a part of the thermohaline circulation of the ocean.

The adjustments which the water masses on either flank of the Gulf Stream must make during the passage of a series of meanders should also be an interesting subject of study. Recording pressure gauges set on the bottom of the ocean, or on sea mounts, might give indications of these motions. Unfortunately, the bathythermographic data on hand are not sufficient for any statistical study of fluctuations of properties in the slope waters and coastal waters.

There is no certainty, of course, that the growth of the meanders observed in the Gulf Stream is necessarily due to an instability of the flow such as is conceived in the model of the two-layer meander theory. The meanders might conceivably be formed and forced to great amplitudes by the wind without the assistance of any instability. In the weather maps for the months of May and June, 1950, there is no marked wind pattern that could be directly forcing the meander pattern observed in the Multiple Ship Survey, and no violent storms over limited areas are noted.

If we consider the eddy formation observed in the course of the Multiple Ship Survey of June, 1950, we note that the Stream is continuous and relatively straight both upstream and downstream of the eddy at 60° W. (fig. 40). One wonders whether there was any marked variation in the component of surface wind stress along the current at 60° W. From the North Atlantic surface wind maps made up at La Guardia Field it is possible to obtain the daily ship reports for every 5° along the 40° N. circle of latitude from 70 to 50° W. If we then compute five-day means of the surface wind stress we find that there was a region of low downstream stress west of 60° W., and a region of fairly high stress east of 60° W. This pattern persisted for almost thirty days before the survey of the eddy. The results are given in table 7, the selected coefficient of shear stress being 2×10^{-3}.

TABLE 7

MEAN COMPONENT OF DAILY WIND STRESS IN DOWNSTREAM DIRECTION
FROM MAY 15, 1950, TO JUNE 19, 1950, ON 40° N. LATITUDE

Longitude	70° W.	65° W.	60° W.	55° W.	50° W.	45° W.
Eastward computed wind stress (dynes/cm.²)	− 0·05	+ 0·01	− 0·06[a]	+ 0·30	+ 0·20	+ 0·40

[a] Position of eddy.

It is natural to wonder whether the coincidence of the eddy with the abrupt change in wind stress is simply a chance coincidence, or expresses a cause-and-effect relationship between the eddy and a monthly mean feature of the wind stress. If the change in wind stress occurs in association with a change in mean current velocity (towed-electrode data on the Multiple Ship Survey of 1950 indicate that east of 60° W. the current velocities fell to about 50 per cent of the velocities to the west), the section of the Stream at 60° W. could act as an important block to passage of meanders. Meanders that were arrested at such a point in the Stream might easily grow into eddies. This is all conjecture, of course; other explanations are possible. For example, in Tolstoy's bathymetric chart (see fig. 16) is a series of sea mounts, some as shallow as 1500 m., in the region where the Gulf Stream reaches 65 to 60° W. longitude. It is conceivable that these sea mounts, and other uncharted ones in the region, are the perturbing cause of the breakdown of the Stream in this area. Bottom features are known to exert an influence upon the path of some surface currents, as is shown, for example, in the case of the Antarctic Circumpolar Current, by Sverdrup, Johnson, and Fleming (1942).

STABLE MEANDERS IN A STREAM OF UNIFORM POTENTIAL VORTICITY

The following dynamical study treats the problem of stable meanders in a Stream of uniform absolute vorticity.

We shall suppose, as before, that absolute vorticity is conserved in the upper layer, and hence that

$$\frac{f}{h_0} = \frac{f + \dfrac{\partial v}{\partial x}}{h}, \tag{9}$$

where h_0 is the value of $h(x)$ at $x \to \infty$.

We shall also suppose that the Stream is not straight, but slightly curved, with a radius of curvature $\mathscr{R}$. If $\mathscr{R}$ is much greater than the width of the Stream, and if we assign a positive sign to $\mathscr{R}$ for cyclonic curvature, then the dynamical equation of the upper layer is

$$fv + \frac{v^2}{\mathscr{R}} = g' \frac{\partial h}{\partial x}, \tag{10}$$

where $g' = g(\Delta\rho/\rho)$. Eliminate h by cross-differentiation; and in this way we obtain

$$v + \frac{v^2}{\alpha \mathscr{R} c} = \frac{1}{\alpha^2} \frac{\partial^2 v}{\partial x^2}, \tag{11}$$

where $c = \sqrt{(g'h_0)}$, and $\alpha = f/c$. We regard the nonlinear term as small in comparison with the others; hence, we express v in series form:

$$v = v_0 + \frac{v_1}{\alpha \mathscr{R}} + \frac{v_2}{(\alpha \mathscr{R})^2} + \dots . \tag{12}$$

Therefore, the functions $v_0, v_1, v_2, \dots$ are determined by the following equations:

$$v_0 = \alpha^{-2} v_0'', \tag{13}$$

$$v_1 + c^{-1} v_0^2 = \alpha^{-2} v_1'', \tag{14}$$

$$v_2 + c^{-1} 2v_0 v_1 = \alpha^{-2} v_2'', \tag{15}$$

$$v_3 + c^{-1}(2v_0 v_2 + v_1^2) = \alpha^{-2} v_3'', \tag{16}$$

the solutions of which are:

$$v_0 = ce^{-\alpha x}, \tag{17}$$

$$v_1 = \frac{c}{3} e^{-2\alpha x}, \tag{18}$$

$$v_2 = \frac{c}{12} e^{-3\alpha x}, \tag{19}$$

$$v_3 = \frac{5c}{18} e^{-4\alpha x}. \tag{20}$$

The first approximation of h may be obtained from equation (10) by direct integration:

$$h = \frac{-1}{g'} \int_x^\infty \left[f\left(v_0 + \frac{v_1}{\alpha \mathscr{R}}\right) + \frac{v_0^2}{\mathscr{R}} \right] dx + h_0$$

$$= h_0 \left(1 - e^{-\alpha x} - \frac{2}{3\alpha \mathscr{R}} e^{-2\alpha x} \right). \tag{21}$$

The value of x for which h vanishes is given approximately by

$$\alpha x \doteq \frac{2}{3\alpha \mathscr{R}}. \tag{22}$$

Because the shape of the interface is not strikingly different in troughs and crests, it is not likely that it would be very easy to test the hypotheses entering into the theory by hydrographic sections across crests and troughs of meanders. A more plausible test is to compare the velocity in a crest and a trough at the same value of h. For purposes of illustration we construct a

TABLE 8

NUMERICAL EXAMPLE OF VELOCITY PROFILE IN MEANDERS

h/h_0	v/c (cyclonic)[a]	v/c (anticyclonic)[a]
0·0.	0·958	1·060
0·1.	0·862	0·950
0·2.	0·772	0·829
0·3.	0·680	0·720
0·4.	0·590	0·620
0·5.	0·495	0·510
0·6.	0·395	0·408
0·7[b].	0·300	0·300
0·8.	0·200	0·200
0·9.	0·100	0·100
1·0.	0·000	0·000

[a] Probable error, ± 0.002.
[b] For h/h_0 values of 0·7 and higher, the values of v/c (cyclonic) and v/c (anticyclonic) are essentially equal, within graphical error.

table of h and v for a meander with radius of curvature $\mathscr{R} = (2/3) \times 10^7$ cm. and $\alpha = 10^{-7}$/cm. The values of v are then plotted against h for both cyclonic and anticyclonic curvature. The results, scaled from the plot, are shown in table 8. The difference is most pronounced for small values of h/h_0. The proper place for a test is probably at about $h/h_0 = 0\cdot2$.

Since the total transport of the Stream in a crest is greater than that through a trough, the meander pattern moves, as a whole, downstream. The rate of this progression may be found by the following simple reasoning. First evaluate the total transport T:

$$T = \int_{x=2/(3\alpha^2\mathscr{R})}^{\infty} vh\,dx$$
$$\doteqdot ch_0\left[\frac{1}{2\alpha} - \frac{1}{6\alpha^2\mathscr{R}}\right]. \tag{23}$$

When the values of parameters introduced above are used, the transport at the crest is greater than that through the trough by about 10 per cent of the average transport. Suppose that the shape of the meander is

$$x = x_0 \cos(ky - vt). \tag{24}$$

At time $t = 0$, the rate of increase of volume W of warm water between $ky = 0$ and $ky = \pi$ due to the motion of the meander is $2x_0 h_0(v/k)$. This, of course, must be equal to the excess of transport through a crest $(ky=0)$ over that through a trough $(ky=\pi)$, $ch_0/3\alpha^2\mathscr{R}$. But at $t=0$ and $y=0$, the radius of curvature $\mathscr{R}$ is also given in terms of the equation

$$\mathscr{R}^{-1} = \frac{\partial^2 x}{\partial y^2} = x_0 k^2; \tag{25}$$

hence

$$\frac{v}{k} = \frac{c^3 k^2}{6f^2}. \tag{26}$$

The rate of progress of a meander with a wavelength of about 300 km. is therefore about 5 cm./sec. The actual movement of meanders ought to be studied extensively by means of the air-borne radiation thermometer.

Chapter Ten

FLUCTUATIONS IN THE CURRENTS

Many catastrophes of an economic kind, such as the failure of the rice crop in Japan, or of a certain fishery, or years of unusual numbers of icebergs in shipping lanes, are attributed to fluctuations in ocean currents. Very little is really known about such fluctuations. It takes years of careful and expensive observation to produce even a very crude description of them. The scientific programs of our oceanographic institutions are not geared to long-term problems of this kind; there is much pressure for novelty, much temptation to follow the latest fad, and a persistent though erroneous notion that all worth-while problems will eventually be solved by some simple, ingenious idea or clever gadget. A well-planned long-term survey designed to reveal fluctuations in ocean currents would be expensive and time-consuming. It might even fail, because of inadequacies of the tools we have at hand. But until this burdensome and not immediately reward-ing task is undertaken, our information about the fluctuations of ocean currents will always be fragmentary.

THE DYNAMICS OF THE FLORIDA CURRENT

Numerous studies of tide-gauge data have been made (Hela, 1952; Mont-gomery, 1938*b*), to obtain information about seasonal fluctuations in the cross-stream slope of the sea surface; and then, by the geostrophic equa-tion, to compute seasonal fluctuations of the mean surface velocity of the Florida Current. In addition to the tide-gauge data already published concerning Key West, Miami, and Cat Key, there are now available, for the first time, tide-gauge data for Havana. This information is obtained from a gauge set up by the United States Coast and Geodetic Survey in 1946. A leveling survey has been made between Key West and Miami (Montgomery, 1941a), but it is, of course, impossible to connect Havana or Cat Key to the same datum by ordinary means (Montgomery, 1947). Therefore we cannot obtain true differences in sea level across the Straits, but we can obtain fluctuations in the differences of sea level at two stations, as shown in fig. 70. At all four stations the maximum sea level occurs in

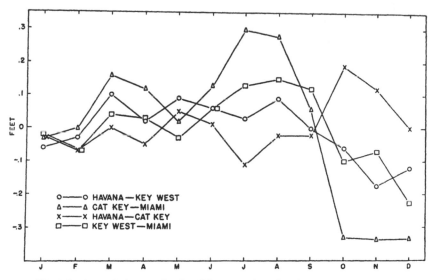

Fig. 70. Annual changes in the difference of sea level between various pairs of stations on the Florida Straits. From Stommel (1953, fig. 1).

September and October, apparently caused by the summer heating of the water. This average rise does not appear in the differences.

One interesting feature of these differences is that the fluctuation in the Miami–Cat Key differences is about twice that of the Key West–Havana differences. The maximum cross-stream slopes occur during July, when the flow is strongest (Hela, 1952), and the maximum downstream slope from Key West to Miami also occurs at this time. These relations of slopes to surface velocity are quite what might be expected from the most elementary considerations of Bernoulli's equation and the geostrophic equation (see Chapter III). But the fact that the fluctuations of the differences at the Miami–Cat Key section are larger than at the Key West–Havana section, and the fact that the downstream slope between Havana and Cat

Key is small in July, rather than largest, are at first sight rather startling and require explanation.

Another noteworthy feature of the Florida Current is that the region of anticyclonic shear in the section off Miami is extensive. So far as direct velocity measurements from an anchored vessel are concerned, the data

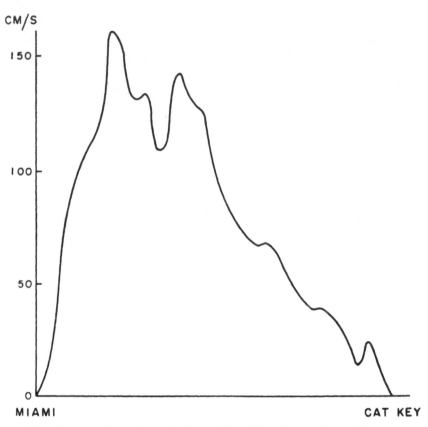

CM/S

Fig. 71. Sample surface velocity profile of the Florida Current between Miami and Cat Key, according to Murray's (1952) towed-electrode measurements.

are scanty. The only observations at different depths are those made by Pillsbury (1891), and these show a rapid decrease of velocity with depth. The axis of maximum surface velocity is not in the center, but is displaced toward Miami; and this feature has been verified on many crossings of the Current off Miami by Murray (1952). Fig. 71 shows measurements made by towed electrodes on one of these crossings, but it should be emphasized that the towed-electrode method is likely to give particularly misleading readings of velocity in the Florida Straits. Nevertheless, the general fact is that a wide zone of anticyclonic vorticity (of approximately $-[0\cdot5\pm0\cdot1]f$, where f is the local Coriolis parameter) seems to be established and also requires explanation. It is, of course, impossible for the variation of the Coriolis parameter with latitude to cause such a shear over so short a distance as the length of the Florida Straits.

At Miami the channel is only about one-half as deep and one-half as wide as at Key West. The change in depth can have no important in-

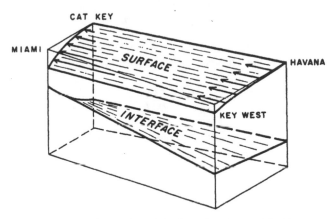

Fig. 72. Hypothetical position of the interface in the Florida Straits. The arrows indicate the magnitude of the current. In the actual Florida Straits there is a right-angle bend in the channel between Key West–Havana and Miami–Cat Key. Bottom topography is not indicated. From Stommel (1953, fig. 4).

fluence on the flow, since the bottom water does not have an appreciable velocity at either section. The narrowing of the channel at Miami is very important hydrographically, for in order to pass through it the water is accelerated, and this requires a small drop in the level of the free surface from Key West to Miami. Since the lower layers are at rest (except for tides), the isopycnic surfaces must slope upward toward Miami to counter-act the axial pressure gradient in the upper accelerating layers. Because this slope must be some 500 times the drop of the free surface, it produces a marked decrease in the thickness of the surface layers as Miami is approached. By conservation of potential vorticity, this vertical shrinking produces an anticyclonic shear and magnifies the transverse geostrophic slope of the free surface. A crude quantitative analysis based on a two-layer model is presented below. Fig. 72 illustrates the average hydrographic structure of the Florida Current envisaged in this hypothesis.

Consider a layer of water of density ρ_1, depth D, overlying a deep, resting layer of density ρ_2. The upper layer has a velocity U at the Key West–Havana section ($x=0$), and its vorticity is zero. Farther downstream ($x=\Delta x$), at the narrower Miami–Cat Key section, the Stream has accelerated by a drop in the free surface Δh. Because the lower layer is at rest, the interface must slope upward to offset the effect of the free surface. If the axial pressure gradient vanishes in the lower layer, then the following relation must hold:

$$\left(\frac{\rho_2-\rho_1}{\rho_2}\right)\frac{\partial D}{\partial x}=\frac{\partial h}{\partial x}.\tag{1}$$

The linearized vorticity equation in this simple case may be written as follows, where ζ is the vorticity of the upper layer:

$$U\frac{\partial}{\partial x}\left(\frac{\zeta+f}{D}\right)=0.\tag{2}$$

We assume that the vorticity vanishes at Key West; hence the vorticity at Miami is given approximately by

$$\Delta\zeta=f\frac{\Delta h}{D}\frac{\rho_2}{\rho_2-\rho_1}.\tag{3}$$

The actual acceleration observed at Miami corresponds to a value of Δh of about -20 cm. (This is somewhat in excess of the $-4\cdot9$ cm. obtained by leveling between Key West and Miami [Montgomery, 1941a].) The density difference of the two layers is approximately $(\rho_2-\rho_1)/\rho_2\cong 2\times10^{-3}$. The initial depth D of the current is roughly 250 m.; and this yields a value of the vorticity increase between Key West and Miami of $\Delta\zeta\cong-0\cdot4\,f$, the observed value given above.

If the high anticyclonic shear at Miami is actually due to the vertical shrinking of the upper layers, it should be possible, by careful hydrographic observations on the two sections, to try to detect the average rise of about 100 m. in the isopycnic surface at a depth of 100–300 m. It would be difficult and tedious to measure this hypothetical rise, because the rise may be partly masked by the large cross-stream slope of the isopycnic surfaces, and by variations in depth caused by the tides. Present data are insufficient to form a basis for a satisfactory test.

The Miami tide gauge is most sensitive to fluctuations in the transport through the Florida Straits, because it lies at the western end of the shallowest section of the Stream.

IRREGULAR FLUCTUATIONS IN TRANSPORT OF
THE FLORIDA CURRENT

The most spectacular fluctuations of the Florida Current are to be seen in
the periods of extraordinary flow revealed by the submarine-cable measurements throughout the months of December, 1952, and January and April,
1953 (fig. 73). During these months the Florida Current maintained a flow
of about 8×10^6 m.³/sec. in excess of its normal flow. Mr Jerome Namias

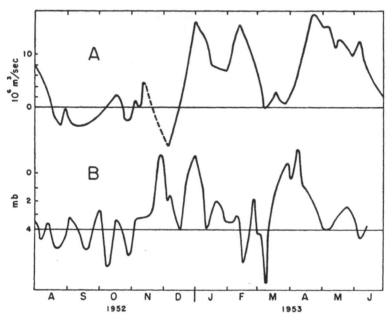

Fig. 73. Comparison of high-flow regimes in the Florida Current, as obtained
from the submarine-cable measurements (curve A), with those computed from
the five-day mean zonal pressure difference between 30 and 20° N. (curve B).
The upper curve (A) shows excess of flow above mean flow.

suggested to me that these periods of excessive flow might be associated
with thirty-day persistent breakdowns of the North Atlantic wind circulation. After a number of trial analyses of weather maps, I decided to use
the five-day mean pressure difference between 30 and 20° N., averaged all
across the ocean, as an index of the strength of the trades north of the
West Indies. These five-day mean zonal pressure differences (30° minus 20°)
are plotted in fig. 73.

It is clear that there are three sustained periods of breakdown of the
Bermuda–Azores high, and that invariably they precede the periods of

high flow of the Florida Current by about a month. It is unlikely that the atmospheric-pressure disturbance is the direct cause of the anomalous flow, for two reasons. First, an adjustment to a pressure change would occur with the speed of a long gravitational surface wave—that is, in several hours, not several weeks. Secondly, the mass flux to compensate the surface-pressure drop would be of the order of 3×10^{11} m.^{3}, whereas the total excess flow of the Florida Current observed during each period is of the order of 3×10^{13} m.^{3}. It seems more likely that the anomalous currents are caused by changes in wind-stress distribution over the ocean. But even this influence is not direct, because the anomalous wind conditions associate *low* winds with *high* discharge of the Florida Current. However, if it is remembered that, because of the immense mass of water involved, the dynamic topography of the North Atlantic subtropical gyre is constrained to remain relatively unaffected by short-period (even thirty-day) wind changes, it is possible to offer a rationalization for this inverse relation in the following manner: All along the Lesser Antilles, at the entrance to the Caribbean Sea, the dynamic topography is such as to produce southwest-flowing currents in the upper 400 m. In the very surface layers the normal trades tend to produce a net northward drift (Montgomery, 1936a). The combined effect of these two tendencies is to produce an essentially west-ward flow of the upper 100 m. at 25° N., and a southerly component below this depth. The fact that Sargasso Sea water normally enters the Caribbean only below 100 m. and not at the surface (Parr, 1939a) lends support to the explanation just outlined.

When the trades cease to blow at 25° N., however, the northward Ekman drift ceases abruptly (within 18 hr.), and even the surface waters move southwestward. Thus the flow into the Caribbean is increased. An order-of-magnitude estimate of this increase may be obtained from the following simple relation. In the case of a steady wind stress τ_x acting in the x-direction (toward the east), the northward Ekman transport rM_{ye} (see Ekman, 1905) is given by the relation

$$rM_{ye} = -\frac{r\tau_x}{f}, \qquad (4)$$

where f is the Coriolis parameter, and r is the width of the ocean. Since τ_x is normally very nearly -1 dyne/cm.^{2} at 25° N. latitude, $f \cong 0.6 \times 10^{-4}$/sec., and $r \cong 5 \times 10^8$ cm., the northward Ekman transport across the 25° N. parallel is 8.3×10^6 m.^{3}/sec. When the trades cease, this flux is added to the water entering the Caribbean. There is a distance of about 3000 km. through the Caribbean to be traversed before this increase reaches the Florida Straits. There is therefore the possibility of a lag of about fifteen days before the first effect of the change in flow appears in the Florida Current.

A good check on these ideas could be obtained by means of potential measurements on a middle-Caribbean cable such as the one between Curaçao and Santo Domingo. The time lag should be about half that observed in the Florida Straits.

It is important to be aware that these remarks about causes of fluctuations of the Florida Current are only speculations and ideas, not proven theories in the usual physical sense.

That such facile speculations may be treacherous is illustrated by certain occurrences that have been observed. In January, 1955, there was an exceptionally intense and widespread month-long breakdown of the Bermuda–Azores atmospheric high-pressure cell, and the northern fringes of the trades disappeared. This meteorological event was accompanied by a decrease in the transport of water through the Florida Straits—a singular instance of complete reversal of the simple explanation discussed above.

SEASONAL FLUCTUATIONS OF THE GULF STREAM NORTHEAST OF CAPE HATTERAS

In the previous chapters in which we discussed the details of single cross sections of the Gulf Stream we were on fairly solid observational ground. In Chapter IX, on meanders, the dangers of picking one of many different possible interpretations of the data, and ignoring the other interpretations, arose. There was an uncomfortable feeling of engaging more with a creation of our own minds than with a fact of nature. As we now proceed to discuss the fluctuations of the Gulf Stream, we are even more apt to find ourselves pursuing a will-o'-the-wisp. Although we do know something about seasonal changes in the shallow wind-drift currents in the upper 100 m. of the ocean, and although we do have fairly adequate data, from tide gauges and the submarine cable, for a study of the Florida Current, we know virtually nothing about changes in the transport of the Gulf Stream itself. With these words of warning, we may now examine what little material there is at hand.

There has never been a cruise which failed to find the Gulf Stream between the United States and Bermuda. It is quite clearly a permanent feature. But are the position and transport of the Gulf Stream entirely independent of the seasonal fluctuations in the strength and position of the wind system?

The pilot charts which are based on hundreds of ship reports on file with the United States Navy Hydrographic Office indicate the presence of seasonal changes in the surface currents. Fuglister (1951a) has prepared a summary of these surface-current data in miles per day. They are presented in fig. 74. The curves for the Florida Current off Miami, the Florida Current southwest of Cape Hatteras, and the Gulf Stream northeast of

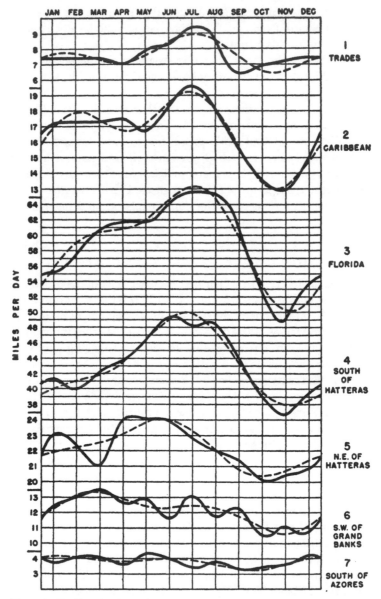

Fig. 74. Fuglister's plots of annual variation of surface currents in various areas of the North Atlantic. From Fuglister (1951a, fig. 2).

145

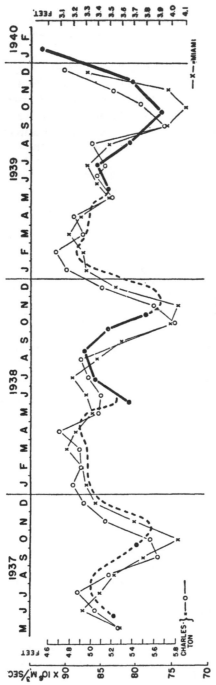

Fig. 75. Comparison of Gulf Stream transport determinations by hydrographic sections (heavy black lines and dots) with those made by tide-gauge records of sea level. From Iselin (1940, fig. 27).

Hatteras, all show a pronounced maximum in early summer. The broken lines represent combined annual and semiannual components obtained by harmonic analysis.

Two other types of data have been advanced to support the idea that there is a seasonal fluctuation in the Gulf Stream. Iselin (1940) computed the volume transport of the Gulf Stream from thirteen hydrographic sections made in the period 1937–1940, and believed that these showed an annual variation of about 15×10^6 m.^{3}/sec. in the transport of the Stream. These data are plotted in heavy black dots in fig. 75.

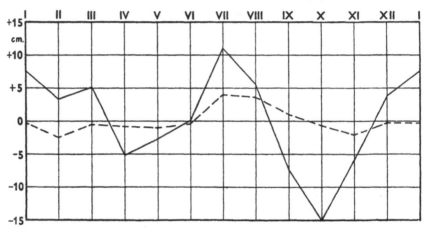

Fig. 76. Annual variations of differences of sea level reduced to normal atmospheric pressure. The solid line is used for the Bermuda–Charleston section, the broken line for that from Key West to Miami. From Montgomery (1938b, fig. 71).

The monthly mean sea level from tide-gauge records should also be an indication of changes in transport.

The geostrophic equation requires that the transverse slope of the free surface and surface velocity be proportional. A decrease of mean sea level at a station along the United States coast suggests an increase in Gulf Stream transport, but does not prove it. Montgomery (1938b) has pointed out a number of corrections (such as allowance for seasonal variations in mean atmospheric pressure) which must be applied to the tide-gauge data.

Records from Miami and Charleston are also plotted in fig. 75. Miami and Charleston are on the same side of the Stream, and hence the records from these two places do not, strictly speaking, serve as an indication of changes in the cross-stream slope. Montgomery (1938b) has plotted a curve for changes in the cross-stream slope between Bermuda and Charleston, reproduced here as fig. 76.

The chief difficulties in utilizing tide-gauge data are: (i) it has been impossible, thus far, to make the proper allowance for the slope of the water surface produced by the direct stress of the local wind over the continental shelf; (ii) the cross-stream slope is related only to surface velocity, and, unless something definite is known about the vertical distribution of velocity, does not give an unambiguous value of the total transport.

One of the most striking features of fig. 75 is the small amount of hydrographic data available, and, worse yet, the concentration of the data toward the latter half of each year. The project was interrupted by the Second World War and was never resumed. The broken-line curve is drawn as a kind of fanciful interpolation, and the ordinate scales of the tide-gauge data are chosen simply to make the curves look alike. I think one is well advised, when confronted by a set of curves such as those in fig. 75, to regard them as merely suggestive, rather than to imagine that they prove anything.

THEORETICAL STUDIES OF TRANSIENT CURRENTS

Several attempts have been made to construct theoretical models that will help in understanding the transient state of the ocean circulation brought about by forced changes in the winds. For example, Ichiye (1951) has obtained an interesting solution for a wind-driven circulation produced by a periodic wind stress acting upon a homogeneous ocean within a rectangular ocean. He uses the frictional boundary layer for the Gulf Stream. He finds that the dynamic topography on the western coast consists of a series of 'waves' of dynamic height moving toward the shore with increasing amplitude at the rate of one crest per year. Thus the annual fluctuation of transport of the Gulf Stream is associated with a shift in the Stream's position. Veronis and Morgan (1955) have obtained a result very different from Ichiye's. The difference apparently arises from their inclusion of the important time-variable term in the mass-continuity equation (which Ichiye takes as vanishing). Veronis and Morgan treat a homogeneous ocean, and find: that the mass transport of the Gulf Stream responds to variable wind systems of more than three months' duration with lags of the order of nine days; and that there is no noticeable shift in the position of the Stream throughout the year.

Both these theoretical studies treat the ocean as though it were homogeneous, whereas, of course, the real ocean is stratified in layers of different density. In the subtropical regions of the Atlantic Ocean, most of the density change with depth occurs in the thermocline, as we have already seen (Chapter IV), and we get some approximation to the true density stratification of the ocean by considering a two-layer ocean model. We must now ask ourselves, what influence does the density stratification of the ocean have upon its response to transient wind systems? Even very crude physical arguments seem to suggest that the density stratification of the ocean cannot respond rapidly to changes in the wind distribution. Examples of such arguments are as follows:

a) There is an immense store of available potential energy in the deep warm-water mass in the Sargasso Sea, more than a thousand times the kinetic energy of all the currents of the North Atlantic. We may suppose that, on the average, the rate of dissipation of the kinetic energy of large-scale currents in the ocean is not more than the rate of work done by the stress of the wind on the ocean. (Much of the work done by the wind may be dissipated in waves or in vertical shear of the surface currents.) If the trade winds exert a stress of 1 dyne/cm.^{2} on the sea surface, and the surface current there is roughly 20 cm./sec., the rate of work done on the ocean is roughly 20 ergs/cm.^{2}/sec., and this sets a maximum average rate to dissipation of kinetic energy per unit area of the ocean currents by turbulence. We next compute the potential energy stored in the warm central core of the North Atlantic Ocean in excess of that which would exist were the thermocline level everywhere. In the latter case the thermocline would be at a uniform depth of about 550 m. In reality, the thermocline is at a depth of 900 m. in half the ocean, and 200 m. in the remainder. If the difference in density of the layers above and below the thermocline is taken as 3×10^{-3} g./cm.^{3}, the potential energy stored in the warm core is 3×10^9 ergs/cm.^{2}. The potential energy stored is thus sufficient to maintain the current system for at least 1700 days in the absence of the driving stress of the winds. The average kinetic energy of the North Atlantic Ocean is of the order of 4×10^6 ergs/cm.^{2}. It is no wonder, then, that the density structure of the Gulf Stream System is so uniform and constant a feature regardless of the vagaries of the weather.

b) An alternative order-of-magnitude computation of the time constant for the decay of the wind-driven circulation following a complete cessation of the wind may be based on the time required for the volume of warm water in the Central North Atlantic Ocean (3000 km. × 6500 km. × 0·5 km.) to drain off at the full Gulf Stream rate of discharge (70×10^6 m.^{3}/sec.), supposing all the Gulf Stream water to be lost to the Arctic seas. This period turns out to be 1600 days.

c) Another estimate of the time constant is provided by a comparison of the total volume of warm water with the rate of convergence of surface water due to the anticyclonic wind system: about 1000 days.

These qualitative considerations suggest that the density structure of the ocean (and hence the current field as usually computed from hydrographic-station data) does not respond completely to those fluctuations in the wind field which have periods of much less than a year. Since such arguments as the ones given under *a–c* might very well be misleading, Dr George Veronis and I (1956) have studied in detail the response of a two-density ocean to a variable applied wind stress. The model studied does not have boundaries, but the applied wind stress is taken as periodic in space. Friction does not appear to be an important factor for periods of a year or less, and inertial-gravitational wave motions are negligible for periods of more than several days. In order that the motion for a very long period shall approach the motion represented in the steady-state solutions by Sverdrup, Reid, and Munk (see Chapter VII), it is necessary to include β, the variation of Coriolis parameter with latitude. For very long periods (thirty years or more), the theoretical currents induced by the wind are confined entirely to the upper layer, as is very nearly true of wind-driven currents of the real ocean, and this surface intensification of currents produced by a transient wind system does not occur without β.

The essential point to make clear is that for periods of between a week and a year the density field in the ocean at mid-latitudes does not attain a full equilibrium response to the variable wind-stress pattern at any instant; but it does respond. Thus, in the North Atlantic, we cannot expect the depth of the main thermocline to adjust itself completely to the seasonal progress of the Bermuda–Azores anticyclonic wind system; on the other hand, the main thermocline cannot be wholly insensitive to the seasonal wind changes, although it would be essentially insensitive to wind-stress changes with periods less than a week. The currents induced by wind-stress patterns with periods between a week and a year are not confined to either layer, but the vertically integrated transport is very nearly in equilibrium with the applied wind stress at any instant. Thus, so far as vertically integrated transport is concerned, the results obtained by Veronis and Morgan (1955) using a homogeneous model and a viscous Gulf Stream should be applicable even for a density-stratified ocean. However, the wind-induced variations of oceanic circulation should *not* be accurately determinable by the traditional method of geostrophic computation from the density field; and the currents are so small that they cannot be observed directly by current meter; and the horizontal displacements involved are too small to produce noticeable changes in the distribution of deep water masses.

Apparently, in central oceanic areas (away from the equator) the chief directly observable effect of the purely wind-induced seasonal variation of oceanic circulation is the change in mean sea level, but even this is nearly obscured by seasonal changes associated with changes of local density arising from local heating, salinity variations, and so on. The island of

Bermuda is favorably placed near the western side of the North Atlantic Ocean, close to the region of maximum mean wind curl, but well away from the Gulf Stream. One should expect to find changes of sea level there which correspond closely to those described in the theory of Veronis and Morgan.

Dr June Pattullo, of the Scripps Institution of Oceanography, has been engaged for several years in making a study of seasonal changes in mean

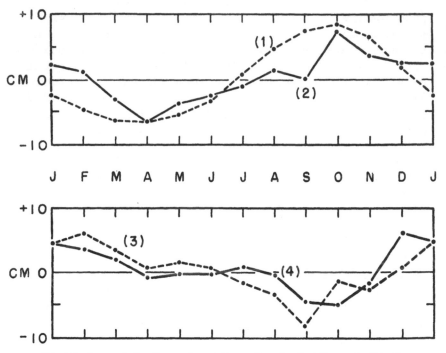

Fig. 77. An analysis of annual variations of sea level at Bermuda. The four curves are described in the text.

sea level throughout the world, and although the results of this valuable project have not yet been published,[1] she very kindly communicated to me the results of her analysis of the Bermuda sea level. Curves 1 and 2 of fig. 77 are plotted from Dr Pattullo's computations. Curve 1 presents the monthly mean sea level computed on the basis of hydrographic-station data alone. It represents the change in sea level due to changes of local density structure and, according to theory, is not directly related to the

[1] Subsequently published by June Pattullo, Walter Munk, Roger Revelle, and Elizabeth Strong, in the *Journal of Marine Research*, 14 (1955):88–156, under the title, 'The Seasonal Oscillation in Sea Level'.

wind-induced circulation. Curve 2 gives the actual tide-gauge data averaged for seventeen years and corrected for atmospheric pressure. The difference between curve 1 and curve 2 is plotted as curve 3.

It is natural to inquire whether the difference (curve 3) can be due to the directly wind-induced changes of oceanic circulation. In order to make a rough, tentative order-of-magnitude calculation, I computed the mean monthly curl of the square of the zonal winds centered at 32°5 N. latitude across the Atlantic. This quantity should be directly proportional to the mean monthly curl of the wind stress in the latitude of Bermuda. According to the theory of Veronis and Morgan (1955), the meridional transport in the central oceanic areas should be given by the relation

$$N = \frac{\text{mean monthly transport}}{\text{mean yearly transport}} = \frac{\text{mean monthly curl of squared zonal winds}}{\text{mean yearly curl of squared zonal winds}}. \quad (5)$$

We shall call this quantity N. I was quite surprised to note that N varies from 0·6 in September and October to a maximum of 1·6 in December. This suggests very much larger variations in oceanic transport than those indicated by Iselin (1940) on the basis of hydrographic-station data. It is important to remember that the changes of transport which we are computing cannot appear in hydrographic data, since they are associated with the barotropic mode. The variation d in sea level due to these monthly variations in transport is given by the formula

$$d = \frac{fT}{gD}(N - 1), \quad (6)$$

where T is the mean yearly transport in the central oceanic area of the North Atlantic at 32°5 N. latitude, and D is the total mean depth of the ocean. In plotting curve 4 I assumed that $f = 0.7 \times 10^{-4}$/sec., $T = 60 \times 10^6$ m.³/sec., and $D = 4$ km. The agreement of this very crude calculation with curve 3 is, to my mind, rather good, and suggests a verification of the theory of transient currents. However, a truly careful discussion of the problem must await the detailed results of Dr Pattullo's studies of the ocean-wide distribution of changes in sea level. There are other physical obstacles to a convincing demonstration. For example, we do not know the ways in which the wind stress depends upon the air–sea temperature difference, and we do not have an adequate theory of the thermohaline circulation in central oceanic areas to use as a companion piece along with the theory of wind-driven circulation. These speculations about the data and about the applicability of the theoretical models will remain inconclusive until really adequate data are available.

Systematic long-duration series of oceanographic measurements are not obtained on a routine basis. The Western Union cable measurements are a solitary example of what is needed. In the future, after unattended instruments on buoys have been maintained for many years at various points in the oceans and as soon as long series of subsurface temperature, salinity, and velocity measurements have accumulated, it will be possible to discuss these matters more satisfactorily.

LONG-PERIOD FLUCTUATIONS IN THE GULF STREAM

There are not sufficient data available to permit us to discuss fluctuations in the Gulf Stream System lasting longer than one year. This is because early measurements were made without a knowledge of the existence of meanders and eddies. Unless allowance can be made for these short-period fluctuations, the long-period variations obtained by plotting the data are not statistically significant.

The Kuroshio, which flows along the coast of Japan, and which appears superficially to be similar to the Gulf Stream, has exhibited a very remarkable change in the period 1919–1950. From 1919 to 1934 (Uda, 1938) the Kuroshio had a pattern of flow past Cape Shiono Misaki very much like that of the Gulf Stream off Cape Hatteras. Between 1935 and 1942 a large, semipermanent cold eddy, 200 km. in diameter, developed at the Cape, and the Kuroshio weakened and went completely around this cold mass. Since 1942 the eddy has slowly decayed and the stream has returned to its original state. We have never observed anything like this in the Gulf Stream.

Chapter Eleven

ROLE OF THE THERMOHALINE CIRCULATION*

Although much effort has been directed toward detailed theoretical elucidation of the consequences of the action of wind stress in producing the oceanic circulation, there has not been a parallel development of the theory of the thermohaline process. The distribution of wind stress over the ocean surface is much better known than the distribution of net heating and cooling of the ocean; thus there is little hope, at present, of constructing as complete and satisfactory a theory of the thermohaline circulation as is possible for the wind-driven circulation. Nevertheless, it does seem desirable to reconnoiter the problem and to try to construct, on a tentative status, a simple theory of the thermohaline circulation that can be applied to real ocean circulations.

In this chapter, attention is first drawn to the fact that the now-familiar 'wind-curl' equation introduced by Sverdrup (1947) is simply a statement of the fact that at every geographical position in the ocean the integrated horizontal divergence of the Ekman wind drift is compensated by a corresponding convergence of the integrated geostrophic current. Next, a thermohaline process is specified in terms of a prescribed vertical velocity structure. At some particular mid-depth the vertical velocity has an extreme value; and this same depth is the level of no horizontal divergence. Two vorticity equations are formed by vertical integration over the layers above and below this level. It is shown that the thermohaline circulation appears as a kind of internal mode of motion, and that the level of no horizontal divergence is also the level of no meridional motion and, if

*For more recent comments on this chapter, see Chapter XIII.

there is a finite interval of no horizontal divergence, this is a level of no motion.

By applying these expressions to the central regions of the oceans, and using the principle of mass conservation, expressions are obtained for the integrated wind-produced and thermohaline-produced transports in the western boundary currents. (It is not necessary to consider the detailed structure of these currents, be they viscous or inertial.) A schematic model of a rectangular basin is examined, and an attempt is made to relate the theory in a more schematic form to actual measurements in the Atlantic Ocean. Several predictions (or more properly, deductions) are made: for example, that there is a great thermohaline countercurrent under the Gulf Stream. It is intended that these predictions shall stimulate direct measurements of deep ocean currents. The well-known discrepancy between the transports of the Gulf Stream and Kuroshio computed from the wind theory (Munk, 1950) and those determined from observation (dynamic computation) is resolved. It is important to note that the Equatorial Current transports in the theories of Sverdrup (1947) and Reid (1948b) do not exhibit this discrepancy. Finally, some of the limitations and perplexities of the rectangular model are discussed.

THE VORTICITY EQUATION FOR THE CENTRAL OCEAN, WIND STRESS ONLY

The physical meaning of the 'curl' equation in the theory of the wind-driven oceanic circulation, as first propounded by Sverdrup (1947) and further developed by Munk (1950), can be most easily appreciated when explained in the following way. We write the linearized stationary equations of motion in the form

$$-f\rho v = -\frac{\partial p}{\partial x} + \frac{\partial \tau_x}{\partial z},$$

$$f\rho u = -\frac{\partial p}{\partial y} + \frac{\partial \tau_y}{\partial z}. \tag{1}$$

Horizontal viscous stresses, inertial terms, and time dependence are neglected. The x-axis is positive toward the east, the y-axis is positive toward the north, and the z-axis is positive upward. The velocity components u and v, the density ρ, the hydrostatic pressure p, and the vertical shearing-stress components τ_x and τ_y are regarded as functions of x, y, and z. The Coriolis parameter f is regarded as a function of y alone.

The equations (1) are now integrated from the surface $z = z_0$ to some great constant depth $z = -h$, at which one supposes that the horizontal pressure gradients and motions vanish. The components of mass transport per unit width are defined in the following manner:

$$M_x = \int_{-h}^{z_0} \rho u \, dz, \quad M_y = \int_{-h}^{z_0} \rho v \, dz, \tag{2}$$

and a function P is introduced:

$$P = \int_{-h}^{z_0} p \, dz. \tag{3}$$

The equations (1) in the integrated form are as follows:

$$-fM_y = -\frac{\partial P}{\partial x} + \tau_x \Big|_{z=z_0},$$
$$fM_x = -\frac{\partial P}{\partial y} + \tau_y \Big|_{z=z_0}. \tag{4}$$

The interchange of integration and differentiation in the pressure terms does not lead to any additional terms, because the lower limit is chosen at zero horizontal pressure gradients, and the terms introduced by the variation in the surface elevation (upper limit), $-p(z_0)\,(\partial z_0/\partial x)$, and so forth, are demonstrably negligible (Munk, 1950). Cross-differentiation of these two equations, and use of the fact that the horizontal divergence of the integrated mass transport vanishes,

$$\frac{\partial M_x}{\partial x} + \frac{\partial M_y}{\partial y} = 0,$$

leads to the 'curl' or vorticity equation now so familiar in theories of wind-driven ocean circulation:

$$\beta M_y = \text{curl} \, \vec{\tau}, \tag{5}$$

where $\vec{\tau}$ is the stress of the wind at the sea surface, and $\beta = \partial f/\partial y$.

In a sense, the simple mathematical manipulation of cross-differentiation tends to obscure the physical meaning of equation (5). Let us therefore begin again with the integrated equations (4) and break each mass-transport component into two parts: one set to represent the Ekman wind-drift transport components M_{xe}, M_{ye}, and the other the geostrophic transport components M_{xg}, M_{yg}:

$$M_x = M_{xe} + M_{xg},$$
$$M_y = M_{ye} + M_{yg}. \tag{6}$$

The equations with which we must deal are the following:

$$-fM_{ye} = \tau_x \big|_{z=z_0}; \quad fM_{xe} = \tau_y \big|_{z=z_0} \tag{7}$$

$$-fM_{yg} = -\frac{\partial P}{\partial x}; \quad fM_{xg} = -\frac{\partial P}{\partial y}. \tag{8}$$

Let us now form the horizontal divergence of the Ekman wind drift, $\mathrm{div}_H \vec{M_e}$, from equations (7):

$$\mathrm{div}_H \vec{M_e} = \frac{\{-\beta M_{ve} + \mathrm{curl}\,\vec{\tau}\}}{f}.\tag{9}$$

This is a quantity which has often been contoured for various oceanic regions in the belief that it has something to do with upwelling.

Because of the variation of the Coriolis parameter with latitude, all northward or southward geostrophic motions exhibit a horizontal divergence, $\mathrm{div}_H \vec{M_g}$, which can be obtained formally from equations (8):

$$\mathrm{div}_H \vec{M_g} = -\frac{\beta M_{vg}}{f}.\tag{10}$$

In a stationary process the total horizontal divergence, $\mathrm{div}_H \vec{M}$, must vanish, hence the sum of equations (9) and (10) vanishes, and we are led back to equation (5). The meaning is clearer, however, because we now see that what equation (5) expresses is the condition under which the divergence of the Ekman wind drift, produced by the wind, is compensated for by the divergence of the geostrophic flow. In Ekman's original picture of the stationary wind-driven currents set up by the wind on a plane ocean with constant Coriolis parameter, the geostrophic flow is nondivergent, and hence he was forced to consider a divergent frictional layer at the bottom to compensate for the surface Ekman wind-drift convergence. Intense lateral friction could produce a divergence, too, but we rule it out by hypothesis. It is worth emphasizing that the role assigned to friction in the present hypothesis is comparatively insignificant: we suppose that it is dynamically important only in the Ekman wind-drift layer. This is a very different point of view from that espoused by Neumann, by Hidaka and the Japanese school, and by Stockmann and the Russian school. As we have seen, it is consistent with the original studies of Sverdrup, Reid, and Munk concerning the interior regions of the ocean; and it is possible that friction (in a climatological-mean sense) is important in certain regions of decay of the intense western currents (for example, in the detaching eddies in the Gulf Stream). The impossibility of stating definitely whether or not friction plays an important dynamical role in the interior of the ocean is one of the reasons why the present discussion is not, strictly speaking, a theory, but merely a hypothesis.

THE VORTICITY EQUATION WITH A
THERMOHALINE PROCESS

Let us now attempt to introduce a simple thermohaline process into this model: For example, we can suppose that, in subtropical latitudes, there is a net vertical flux of heat through the surface of the sea tending to increase the temperature, and hence to decrease the density of the surface waters, and that the density field is actually maintained constant, in spite of this net heat flux, by virtue of a slow vertical mass flux ρw bringing deep water from below into the surface layer. The vertical velocity w is certainly very small at the surface and at the bottom. At some intermediate depth $z = z_i$ it reaches an extreme value (maximum in the subtropics). Since, by mass continuity,

$$\frac{\partial}{\partial x}(\rho u) + \frac{\partial}{\partial y}(\rho v) = -\frac{\partial}{\partial z}(\rho w), \tag{11}$$

the level of extreme vertical velocity, $z = z_i$, also corresponds to the level of no horizontal divergence. In the layers above and below this level there are net divergences, and we shall suppose that they are balanced by geostrophic divergences in these layers. We now divide the ocean into two such layers, indicating the layer above the level of no horizontal divergence by the subscript 1 and the layer below by the subscript 2. We then proceed to form the integrated vorticity equations for each layer. First, the following definitions are introduced:

$$M_{x1} = \int_{z_i}^{0} \rho u \, dz; \quad M_{y1} = \int_{z_i}^{0} \rho v \, dz$$
$$M_{x2} = \int_{D}^{z_i} \rho u \, dz; \quad M_{y2} = \int_{D}^{z_i} \rho v \, dz \tag{12}$$

where D is the depth of the bottom of the ocean. We cross-differentiate the equations of motion (1), obtaining

$$\beta \rho v + f\left[\frac{\partial}{\partial x}(\rho u) + \frac{\partial}{\partial y}(\rho v)\right] = \frac{\partial}{\partial z}\left(\frac{\partial \tau_y}{\partial x} - \frac{\partial \tau_x}{\partial y}\right). \tag{13}$$

If we now assume that vertical shearing stresses are not important below the depth of the Ekman spiral we may take the mid-depth flow as being essentially frictionless at $z = z_i$. Furthermore, for simplicity, we shall take the depth of the bottom, $z = D$, as a constant, in order to avoid, at this stage, the rather superfluous complications of carrying terms of the form $(\rho u)|_{z=D} (\partial D/\partial x)$, and so forth. The integrated forms of the vorticity equations of the two layers are, then:

$$\beta M_{y1} = -f(\rho w)_i + \text{curl} \, \overrightarrow{\tau}, \tag{14}$$

$$\beta M_{y2} = f(\rho w)_i. \tag{15}$$

These equations are of the same form as equation (5), but contain an additional term because of the vertical mass flux associated with the thermohaline circulation. The sum of the two equations is exactly the same as equation (5); and hence we see that the introduction of the thermohaline circulation in this manner does not produce any net transport over both layers in the interior of the ocean, but does result in a kind of internal mode of circulation. As we shall see, there is every reason to suppose that the flow induced in each layer may be of the same order of magnitude as that produced by the wind stress. Since there is so much uncertainty in choosing the depth of no motion for a reference level in dynamic computations from actual hydrographic data, it is important to form a clear idea of the role of this internal thermohaline mode. Indeed, one of the theses of this study is that the apparent discrepancy between the theoretical Gulf Stream transport and that computed 'dynamically' (Munk, 1950) is simply a result of a confusion of this sort. In a later section this will be considered in some detail.

In order to obtain some idea of the magnitude of the velocity w_i required to produce a transport equal to that produced by wind stress, we note that Munk (1950) gives the value $\text{curl} \, \vec{\tau} = -0.7 \times 10^{-8}$ g. cm.$^{-2}$ sec.$^{-2}$ at 35° N. latitude in the North Atlantic Ocean. An upward flux of $(\rho w)_i = 0.85 \times 10^{-4}$ cm./sec. (roughly 8.5 cm./day or 30 m./year) produces an equivalent transport. This is not an exorbitantly high vertical velocity. It corresponds closely to values premised by Riley (1951). It may be compared with what little is known of global heat-budget considerations. For example, suppose that in subtropical regions there is a net downward heat flux of 150 g.cal./day/cm.^{2} into the sea surface, and that this heat is used to heat water flowing up from below. The average temperature increase is in the neighborhood of 15° C., hence the vertical flow is 10 cm./day. These figures should be regarded as maximum heat fluxes near the equator: they would require that the ocean transport more of the equatorial heat surplus toward the poles than does the atmosphere.

From the point of view of equation (13) the level of no horizontal divergence is *also the level of no meridional motion*. Moreover, if there is a small but finite interval of depth in which there is no horizontal divergence, this interval will also be a layer of *no motion* and *no vertical shear of geostrophic motion*. This may be in part an explanation for Defant's (1941) intuitive choice of the level of no motion in the Atlantic Ocean. In all that follows I have been forced to make the assumption that the level of no horizontal divergence is of finite thickness, in order to make it a level of no motion as well.

It is perhaps worth while to reëmphasize at this point that all we have done in this section is to introduce new fields of horizontal divergence associated with a hypothetical thermohaline process, in addition to an Ekman wind-drift divergence at the top of the ocean. These additional divergences are compensated for in each layer by the divergence of the meridional geostrophic flow.

THE TRANSPORT OF THE WESTERN CURRENTS

The role of the western currents in the ocean is to maintain continuity of mass in the ocean basins bounded by coasts. Generally, the solutions of the interior vorticity equations lead to an accumulation of water at certain latitudes and in certain layers, and conservation of mass can only be restored by rapid western currents in which inertial or viscous forces become of dominating importance. In order to compute the transports in the upper, G_1, and lower, G_2, layers of the western currents, we need merely make use of the interior solution for the north transport component and the equation of mass continuity (11) in its integrated forms:

$$\frac{\partial}{\partial x} M_{x1} + \frac{\partial}{\partial y} M_{y1} = (\rho w)_i, \tag{16}$$

$$\frac{\partial}{\partial x} M_{x2} + \frac{\partial}{\partial y} M_{y2} = -(\rho w)_i. \tag{17}$$

Terms from the differentiation of the limits of integration do not appear, since $z = z_i$ is a level of no motion. Let us now consider an ocean bounded at $y = 0$ by the x-axis and by meridional walls at $x = 0, r$. The boundary condition at $x = r$ is taken as $M_{x1} = M_{x2} = 0$, following Sverdrup (1947). The quantities M_{y1} and M_{y2} are known from equations (15), and hence we may find M_{x1}, M_{x2}:

$$M_{x1}(x, y) = \int_x^r \left[-\frac{\partial}{\partial y} M_{y1} + \rho w_i \right] dx, \tag{18}$$

$$M_{x2}(x, y) = \int_x^r \left[-\frac{\partial}{\partial y} M_{y2} - \rho w_i \right] dx. \tag{19}$$

In general we will be led to the result $M_{x1} \neq 0$ and $M_{x2} \neq 0$ at $x = 0$. These fluxes must be regarded as absorbed by the boundary layer or western current solution, hence the western current transports are given by

$$G_1(y) = I_1(y) + \int_0^r \int_0^y \rho w \, dy \, dx, \tag{20}$$

$$G_2(y) = I_2(y) - \int_0^r \int_0^y \rho w \, dy \, dx, \tag{21}$$

where

$$I_1(y) = \frac{1}{\beta} \int_0^r (f\rho w - \text{curl } \vec{\tau}) \, dx, \qquad (22)$$

$$I_2(y) = \frac{1}{\beta} \int_0^r f\rho w \, dx. \qquad (23)$$

WIND-DRIVEN AND THERMOHALINE CIRCULATIONS IN A RECTANGULAR OCEAN

In order to portray some of the chief features of the thermohaline circulation, the analytic solution for a simple rectangular basin has been obtained. Coasts are placed at $x = 0$, $x = r$, and $y = 0$, but the ocean is left open

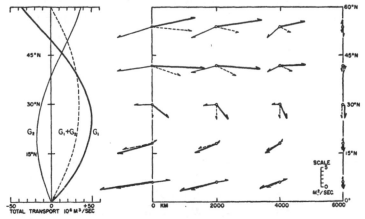

Fig. 78. A rectangular ocean with an idealized thermohaline process. The vertically integrated transports per unit width in the interior (or central) regions of the ocean are shown on the x, y plane at the right of the figure. The broken-line arrows show magnitude and direction of the vertically integrated transport per unit width due to wind stress alone. When the idealized vertical mass flux is introduced, new patterns of flow appear; the heavy solid arrows indicate the transport per unit width of the upper layer, the fine solid arrows indicate that of the lower layer. To the left are shown the total transport functions of the western currents.

on the northern edge. A wind stress of the form $\tau_x = -\tau_0 \cos ly$ is applied to the surface, and a vertical flux of $\rho w_i = \rho w_0 \cos ly$ is fixed at the level of no horizontal divergence. The values used in computing the transports exhibited in fig. 78 were: $r = \pi/l = 6000$ km.; $\tau_0 = 3$ dynes/cm.[2] (this fairly large figure for the surface stress is used to bring the curl at $y = \pi/2l$ up to Munk's value of $0 \cdot 7 \times 10^{-8}$ g./sec./sec. and to include the $1 \cdot 3$ meridional factor); $w_0 = 10^{-4}$ cm./sec.; $\beta = 2 \times 10^{-13}$/cm./sec.; and $f = \beta y$. Detail of the western currents is not shown, but the total transport of the western current in each layer is shown as a function of latitude by the curves on the left-hand side of the diagram. The features of the circulation in this idealized basin, and with this very arbitrary choice of distribution of vertical velocity, cannot, of course, be applied directly to any real oceanic basin, but they do give an indication of the way in which the introduction of the thermohaline process modifies the current pattern in the central ocean regions. The broken-line arrows indicate direction and amount of transport in the upper layer due to wind stress alone, which are similar to those given by the Munk theory. They also represent the total transport of both layers in the thermohaline model, because, as we have seen above, the transports of the thermohaline process are an internal mode of motion, in which the transport of one layer is compensated by a counterflow in the other layer. On the left-hand side of the figure the transport of the upper layer due to wind alone, or alternatively the transport of both layers $(G_1 + G_2)$ in the thermohaline case, in the western current, is indicated by the broken-line curve. Because of the choice of parameters, the maximum transport, 36×10^8 m.^{3}/sec., matches that computed by Munk (1950) for the Gulf Stream.

The introduction of vertical mass flux at an arbitrary level of no horizontal divergence completely alters the picture of the transports, so far as individual layers are concerned. In the upper layer (heavy solid arrows and curves), this modification is revealed as: (i) a general strengthening of transport over the southern half of the basin, (ii) a turning toward the north in the northern half, (iii) an increase in the transport in the western current in low latitudes, (iv) an abrupt cutoff of the western current somewhat south of the maximum westerlies, and (v) a reversal of the western current in the northern quarter of the ocean, contrary to the wind-driven current.

In the lower layer (light solid arrows and curves): (i) strong southwestward transports occur in the northern half of the basin; (ii) this flow concentrates toward the west into a powerful countercurrent underneath the surface western current; (iii) at the latitude of no vertical flux there is no meridional transport in the interior of the ocean, and thus the northern half of the lower layer is completely cut off from the southern half over the entire interior region of the ocean, except for the connection through the counterflow underneath the western current; (iv) the flow in the southern half is northeastward and slow; and (v) the water in the lower layer, in the southern half, is relatively stagnant, and more than 200 years old (that is, since contact with the surface).

It is interesting to note that the behavior of the thermohaline circulation implied by this model in the southern half of the interior of the ocean is inverse to the action of a simple Hadley-type convective circulation: warmed water moves equatorward, cold water poleward. The northern half of the interior solution behaves like a Hadley cell.

TRANSPORT CALCULATIONS FROM HYDROGRAPHIC
DATA FROM THE ATLANTIC OCEAN

The rectangular ocean basin discussed in the above section is not meant to
be a true representation of any real ocean basin. It does, however, illustrate
certain dynamical principles which we will now try to use in a rough,
tentative way in a discussion of the circulation of the North Atlantic Ocean.
In order to make the discussion as quantitative as possible, the geostrophic
transports across various sections in the Atlantic are computed, and an
elementary method for shifting the level of no motion is introduced. There
is, of course, no positively certain way to determine the depth of no motion.
A variety of arguments will therefore be introduced in this connection, none
of which can be regarded as really satisfactory.

In the past two years Worthington has embarked upon a program of
repeating all old (1930–1940) *Atlantis* hydrographic sections. Moreover,
he has made his casts to the very bottom, and thus there is now becoming
available, for the first time, a thorough deep network of sections across the
western North Atlantic from which an adequate study of the geostrophic
transport can be made. As Worthington has pointed out in a recent lecture
at Honolulu, the feature of principal dynamical interest which these new
sections reveal is that there are horizontal density gradients under the Gulf
Stream at all depths sufficient to introduce sizable contributions to geostrophic transports as determined by standard 'dynamical' computation.

As a result, the net geostrophic transport computed for each section
depends strongly upon the choice of the level of no motion, and a change of
this level—for example, from 2000 m. to the bottom at 5000 m.—makes a
very large change in the net transport. In other words, the suppositions
introduced in making the integrations leading to equation (4), namely, that
the horizontal pressure gradients and motions vanish for all depths beneath
a certain great depth, are not valid. Nevertheless, in any particular section
there may be a depth *at which* (though not below which) horizontal pressure
gradients parallel to the section and motions normal to the section vanish.

This uncertainty in the depth of no motion makes it much more difficult
to compare the results of theory and observation than might be inferred
from Iselin's analysis of early observations (1936, 1940). Thus Munk's (1950)
comparison of his own theoretically deduced Gulf Stream transport of
35×10^6 m.^{3}/sec. with the 'observed' transport of 74×10^6 m.^{3}/sec. is fairly
meaningless. The latter figure was computed by Iselin with an assumed
depth of no motion at 2000 m., neglecting counterflows above 2000 m. on
the sides of the main current, and also neglecting counterflows beneath the
Stream. An even worse discrepancy can be obtained, for example, by using
Worthington's new Gulf Stream sections, excluding countercurrents, and

taking the depth of no motion at 5000 m. In this case the net 'observed' transport is 123×10^6 m.³/sec. (Worthington, 1954b). A consistent use of the bottom as the depth of no motion has caused Worthington serious trouble with the conservation of mass, and he has hoped to circumvent this trouble by supposing that there are large-scale nongeostrophic flows in the ocean, particularly between Bermuda and the West Indies. I cannot offer any positive proof to refute Worthington's position on this question, but it is my opinion[1] that these difficulties may be mostly resolved in terms of the thermohaline circulation, that the depth of no motion in the western North Atlantic is, roughly, 1600 m. (in agreement with the estimate by Defant, 1941), and that the actual net transport of the Gulf Stream System is much more nearly equal to Munk's theoretical value of 35×10^6 m.³/sec. than previous estimates neglecting the deep countercurrent underneath the Gulf Stream have indicated. A crucial test of this explanation will be to observe by direct means the direction of flow at various depths beneath the Gulf Stream. A deep counterflow is a necessary, but not sufficient, feature to validate this theory. I therefore emphasize the point that this hypothesis of the thermohaline circulation can be struck a mortal blow should direct observation disprove the existence of a deep countercurrent on the western side of the North Atlantic Ocean.[2]

In order to obtain a rough idea of what the transports in the actual Gulf Stream might be, the following crude analysis is presented.

[1] I have tried to adhere, as much as possible, and perhaps rather slavishly, to the notion that the flow in the deep water is steady and geostrophic, and that the observed data which we have are not much distorted by purely local effects or short-period fluctuations. Recently other students of the subject have begun to doubt both of these assumptions, but the degree to which they are true is not yet known. Even granting these assumptions, there are other possible interpretations of the data which are quite different from the one I describe here. The conception of the current systems presented here is not a unique solution, nor, indeed, are the schemes proposed by other writers. I think the multiplicity of possible interpretations needs emphasis in order to serve as a warning to the reader. The reason I have used such simple assumptions is that otherwise there are so many degrees of freedom that almost any interpretation becomes possible.

[2] As the manuscript for this book was completed in mid-1955, it has not been possible to include more recent developments in the text; however, in March and early April of 1957 a preliminary set of direct measurements of currents under the Gulf Stream (using Swallow's new neutrally buoyant floats) has been made by J. C. Swallow and L. V. Worthington and reported in a letter to *Nature* magazine 179 (June 8, 1957):1183–1184, and since these measurements have an important bearing upon the opinions expressed in the text, I insert an extract from the letter as a footnote added in proof:

'In choosing the most suitable part of the Gulf Stream system in which to work, various factors were considered. The surface velocities in the Stream off the American continent usually exceed 200 cm./sec., and it was felt that such strong currents would

A HYPOTHETICAL ANALYSIS OF THE GEOSTROPHIC
TRANSPORTS IN THE ATLANTIC OCEAN

Let us define a function $T(z, d)$, the transport per unit depth across a hydrographic section, as a function of depth, z, and of the reference level (level of no motion), $z = d$. If $S(z)$ is the ratio of the width of the section at depth z to that at the surface, taken as completely determined by bathymetry, then the transport per unit depth, $T(z, d)$, can be expressed in terms of the same function referred to the bottom $z = b$ as the level of no motion in the following manner:

$$T(z, d) = T(z, b) - \frac{T(d, b)}{S(d)} S(z). \tag{24}$$

The total vertically integrated transport across the section, $\mathcal{T}(d)$, is a function of reference level only:

$$\mathcal{T}(d) = \int_b^0 T(z, d)\, dz. \tag{25}$$

make it difficult to keep track of the deep floats. At Stommel's suggestion, an area off Cape Romain, South Carolina, was chosen. Here the shallow (less than 800 m.) Florida Current flows over the Blake Plateau, while strong pressure gradients are found in the deep water farther offshore. Farther north, towards Cape Hatteras, the violent shallow gradients are superimposed on the deep ones. To the south the deep gradients dwindle away in a manner not yet understood. On these counts it was decided to place the floats as close as was practicable to the junction of the surface Stream with the deep water.

'The bulk of the work was carried out in about lat. 33° N., and between long. 75° 30′ W. and 76° W. Excellent Loran coverage exists in this area. Two ships took part in this work, which was a joint venture of the National Institute of Oceanography and the Woods Hole Oceanographic Institution. The current measurements were made by the R.R.S. DISCOVERY II, while the R.V. ATLANTIS occupied hydrographic stations, in which serial observations of temperature and salinity were made in order to provide a nearly synoptic picture of the deep pressure conditions.

'Nine floats were followed, of which seven were in deep south-going water. The measurements lasted for periods of 1–4 days, with some overlaps when more than one float was being followed. Three floats at 2,500 metres moved in directions between south and south-west with mean velocities between 2·6 and 9·5 cm./sec., and four floats at 2,800 metres depth moved almost due south with velocities of 9·7–17·4 cm./sec. Additional evidence for a south-going deep current was obtained by A. S. Laughton, who took underwater photographs of the deflexion of a ball suspended on a string, only 50 cm. above the sea floor, in a depth of 3,200 metres. A southward movement of about 5 cm./sec. was found at that depth.

'The ATLANTIS hydrographic stations were made at right angles to the drift of each float. The spacing of the stations, eighty-eight in number, did not exceed ten miles in the vicinity of the floats. In one case a rectangular pattern of eight stations not more than three miles apart was laid. There are some irregularities in the slopes of the deep isobars; but the average conditions suggest that the level of no motion in this area most probably lies between 1,500 and 2,000 metres, if the southerly movement of the deep water is to be accounted for by the geostrophic equation.'

In most sections the function $S(z)$ is nearly unity, except in the vicinity of the bottom of the section; thus a graph of $T(z, d)$ for any particular choice of reference level $z=d$ can be approximately transformed to the function $T(z, d')$ for another reference level $z=d'$ by simply shifting the origin of function by a constant amount,

$$T(z, d') \cong T(z, d) - T(d', d),\qquad (26)$$

so that the level of no motion can be made to move up and down without changing the shape of the curve. Thus, in fig. 79, the approximate consequences of a change of reference level on the vertical distribution of transport per unit depth can easily be visualized by shifting the origin of the abscissa to left or right so that the curve will intersect the ordinate at the desired reference level.

In fig. 79, one of the sections at which we can most certainly assign a reference level is section 9 across the Florida Straits. We know from the low salinity of the water at the depth of 600–700 m. immediately north of the Florida Straits that the reference level must be at the bottom, or below it. Direct current measurements and the results of electrical-potential recordings between Key West and Havana indicate that the mean flow is nearly 26×10^6 m.³/sec.; hence, unless the level of no motion is just about at 700 m. (the bottom) a serious discrepancy would exist between this estimate and the vertically integrated geostrophic transport through the section. We therefore take the transport-per-unit-depth curve for section 9, shown in fig. 79, as a starting point in our analysis of the geostrophic transport of the Gulf Stream in particular and the Atlantic Ocean in general. If the flow in the western North Atlantic is geostrophic, then the total mass transport through sections 8 and 9 must be very nearly the same as that through section 6 or section 7. So far as I can determine, this requirement is best met by the choice of reference levels indicated in fig. 79. In order to demonstrate the degree to which the total mass conservation is achieved by this choice, I have prepared the more detailed graph shown here as fig. 80, using some of Worthington's newest data.

It is reassuring to note that not only do the total transports $\mathcal{T}(d)$ balance approximately, but also the transports at each depth $T(z, d)$ balance, very roughly, when the depth of no motion is chosen in the interval of 1200–1600 m. Evidently we cannot expect precise agreement. The shifting of the level of no motion uniformly over the whole section may be too crude; there may be differences in the level of no motion at various points across each section. In any case, the choice of the interval of 1200–1600 m. leads to much smaller violations of mass continuity than the choice of the bottom as reference level.

We can extend the argument of total mass conservation to other sections, but, because of vertical flow, we cannot expect the individual transport per unit depth to be conserved over large areas. Thus, for example, we may reasonably expect the vertically integrated transport to the north across section 7 to be equal to that to the south across section 3, because the Atlantic Ocean is essentially closed toward the north.

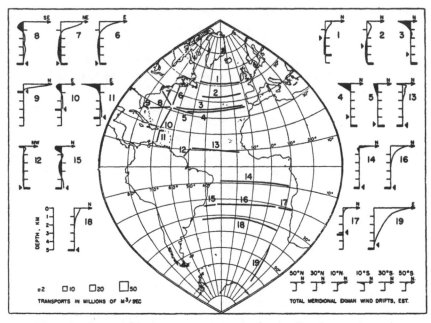

Fig. 79. Geostrophic transport per unit depth, $T(z, d)$, across various sections of the Atlantic Ocean. The total transport across various latitude circles due to Ekman wind drift alone is indicated at the lower right corner; the depth distribution is estimated. Western boundary currents are the major features in sections 6, 7, 8, 9, 10, 11, 12, and 15. The other sections represent interior (or central) oceanic regions.

The broad arrow-point symbol indicates the depth to which actual data extend, below which the curve may be extrapolated; the long, flat rectangle indicates the depth of the solid bottom. The stations used in the respective sections were as follows (A = *Atlantis*; C = *Caryn*; Ch = *Challenger*, 1932 cruises; D = *Discovery*; and M = *Meteor*):

Sec.	Sta.	Sec.	Sta.	Sec.	Sta.
1	Ch 1, 12	8	A 5313, 5328	14	M 188, 201
2	Ch 1287, 1293	9	A 5155, 5162	15	M 165, 167
3	A 4628, 4637	10	C 202, 218	16	M 168, 180
4	A 5307; MVE-3	11	A 5232, 5234,	17	M 181, 183,
5	A 5203, 5210		5235, 5262		184
6	A 5180, 5202	12	M 289, 293	18	M 75, 87
7	A 5295, 5312	13	M 218, 289	19	D 1156, 1165

Sections 3, 4, and 5 are horse-latitude sections where winds are small but the curl of the wind stress is at a maximum. They include parts of the interior that are well removed from the western boundary currents. This is an important region of the ocean from a theoretical point of view. We should establish here whether the interior solution really does correspond to what is expected from the theory of wind-driven currents. It is not an easy section to construct with confidence: no single section is available across the interior of the North Atlantic with many deep-water data (the *Dana* depth determinations are unreliable). Moreover, two stations such as those used in section 3 are really not sufficient below the Mid-Atlantic Ridge: the pressure gradients in the two basins may be different. Section 4 is composite. Below 2000 m. it is most unsatisfactory. Section 5 is made from Worthington's reliable new deep-water data, but it does not include the entire width of the North Atlantic. It is computed from the bottom. The conservation of mass requires that the reference level be placed at about 1200 m., and it is interesting to note that this depth roughly coincides with the level at which the great salt pall from the Mediterranean seems to hang. From the fact that this highly saline water extends almost due westward in a tongue issuing from the Straits of Gibraltar it may be shown that at depths of between 1000 and 2000 m. the value of the meridional component of $T'(z, b)$ must be less than 0.2×10^6 m.^{3}/100 m./sec.

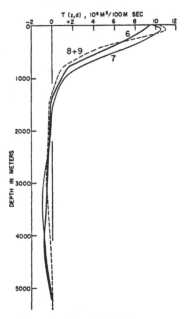

Fig. 80. Detail of the transport-per-unit-depth function $T(z, d)$ for the Gulf Stream sections shown in fig. 79.

Were it greater, the wedge of Mediterranean Water would not point due west—there would be a perceptible tilt. A choice of about 1600 m. is indicated by the curves in fig. 79. In section 5, for which good deep-water data are available, this does not imply large transports of deep water and bottom water. The total vertically integrated transport across section 3 appears to be of the order of

$$30\text{–}40 \times 10^6 \text{ m.}^3/\text{sec.,}$$

a figure not inconsistent with the interior (central) solution given by the Munk theory.

Let us now proceed to discuss other sections in fig. 79. The northernmost

section, section 1, goes from shore to shore, and hence should include interior
and boundary currents. The 1932 *Challenger* stations used extended only
to a depth of 2000 m. If 2000 m. is used as the depth of no motion there is a
net flux above 2000 m. of roughly 30×10^6 m.^{3}/sec. to the north across
section 1. By shifting the level of no motion to about 1500 m., or even
shallower, it is possible to obtain a balance of mass flux across the section
(as depicted in fig. 79). The flux due to Ekman wind drift across section 1 is
shown, approximately, at the bottom of fig. 79. It is too small to play any
important role in the net balance across section 1. Our reasoning forces us
to admit a deep southward flow across section 1 between 1500 m. and the
bottom. A similar pattern of flow appears to occur across section 2, but,
although the data for this section extend to 3000 m., the section does not
extend all the way to the western coast.

In terms of the foregoing hypotheses, the deep currents, which at sections
1 and 2 are distributed over much of the interior, are compressed into a
narrow western boundary current at sections 6 and 7 and do not exist in the
interior at sections 3, 4, and 5. Because of the shallow sill in the Florida
Straits (indeed, even over the Blake Plateau), the deep countercurrent
shown below 1600 m. on sections 6 and 7 splits away (sections 8 and 10)
from that part of the surface current (sections 9 and 11) which passes
through the Caribbean. Because the curl of the wind stress is nearly uniform
between 20 and 40° N., the surface currents across sections 9 and 11 are
more properly regarded as wind-driven than are the surface currents across
sections 8 and 10. By hypothesis, the schematic model of the rectangular
ocean described above causes this deep current to be absorbed by vertical
motion into the surface layer of the interior of the subtropical North
Atlantic. In the real ocean it appears that only part of the deep flow across
sections 6, 7, 8, and 10 can be so absorbed, and that most of it must
flow across the equator into the South Atlantic Ocean, where it appears on
both section 12, beneath the Guiana Current, and section 15, beneath the
Brazil Current. The deep currents of the South Atlantic are mostly confined
to narrow western boundary currents. Sections 12, 13, and 14 are within
10° of the equator, where there is some reason to be slightly suspicious of the
geostrophic calculations. Also, the flow of the Ekman wind drift cannot be
ignored at these latitudes. It is interesting to note that the flow of Antarctic
Intermediate Water appears to be stronger across section 13 than across
section 12; one wonders whether this indicates vertical velocity near 1000 m.
or a surprisingly deep influence of curl of the wind stress. The transport
across interior section 14 seems remarkably low, especially when one notes
by what a narrow margin the geostrophic transport misses being canceled
by the Ekman wind drift.

We now come to sections 15 and 16, which are the counterpart in the

South Atlantic of sections 7 and 3 in the North Atlantic.[3] Sections 3 and 16 are both situated in subtropical regions of high atmospheric pressure: the regions of maximum curl of the mean wind stress between the mid-latitude westerlies and the trades. Sections 3 and 16 show about the same equatorward transport; that across section 16 is, perhaps, a little deeper. Each is very much what would be expected from the wind-stress theory alone. It is only when we compare the two western current sections 7 and 15 that we see a striking difference between hemispheres. Although section 15 is very complicated, the depth of the level of no motion may be ascertained fairly well from water-mass considerations. The choice shown in fig. 79 indicates three such levels between different water masses. It is consistent with the descriptions by previous investigators. The difference in appearance between section 7 and section 15 is so remarkable as to call for further description and elaboration. In both sections the total vertically integrated transport poleward must be equal to the total wind drift equatorward across sections 3 and 16 respectively, from mass-conservation arguments. The wind-driven component of section 7 and that of section 15 must be essentially the same. The marked difference between the function for the Gulf Stream and that of the Brazil Current must therefore be due to the different way in which the thermohaline western boundary currents enter. In the Gulf Stream the thermohaline circulation reënforces the poleward surface motion; in the Brazil Current the situation is quite the contrary. The separation of the wind and thermohaline contributions is shown schematically in fig. 81.

In summary, we see, therefore, that the wind-driven component of the circulation is much the same in the two hemispheres, but that in the western boundary currents the thermohaline circulation plays quite different roles in the Gulf Stream and in the Brazil Current, making the former appear much more strongly on charts of the topography of surfaces, equal pressure, or density.

For purposes of comparison, the function $T(z, d)$ is drawn also for the transport through a north-south *Discovery* section across the Antarctic Circumpolar Current, section 19, 4500 m. being taken as the reference level. The trifling Benguela Current is shown in section 17.

Transport curves for a rough two-layer division of the circulation of the Atlantic Ocean are indicated in fig. 82, a and b. Each line represents about 10×10^6 m.[3]/sec.

[3] Since this chapter was written, a much more precise analysis of deep water transports in the South Atlantic Ocean has been published by Georg Wüst: in Papers in Marine Biology and Oceanography, Supplement to Vol. 3 of *Deep-Sea Research*, 1955, entitled 'Stromgeschwindigkeiten im Tiefen- und Bodenwasser des Atlantischen Ozeans auf Grund dynamischer Berechnung der *Meteor*-Profile der Deutschen Atlantischen Expedition 1925/27', pp. 373–397.

WEAKNESSES AND DIFFICULTIES IN THE
PRESENT THERMOHALINE HYPOTHESIS

The chief weakness of the treatment outlined here, as I see it, is the hypothetical nature of the quantity ρw_i. How shall we measure it directly? If frictional stresses in deep water are really negligible, then much information concerning the field of w might be obtained from an analysis of central oceanic hydrographic data, since

$$\rho \frac{\partial^2 w}{\partial z^2} \doteq \frac{\beta}{f} \frac{\partial \rho v}{\partial z} \doteq \frac{\beta g}{f^2} \frac{\partial \rho}{\partial x}, \tag{27}$$

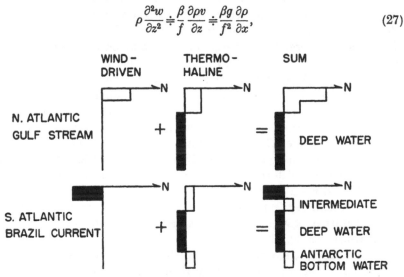

Fig. 81. Schematic diagram of the possible explanation of the very different transport-per-unit-depth curves for the Gulf Stream and the Brazil Current.

but the available very deep data in the North Atlantic are quite insufficient for such a study.

Bottom topography must be of considerable importance in the thermohaline circulation, in which horizontal pressure gradients do not vanish at the bottom. Major topographic features like the Mid-Atlantic Ridge must be of great importance; along the eastern rise of this ridge we might very well find a deep southward current (of the boundary-layer type) similar to the one under the Gulf Stream. Where bottom water is confined to deep basins the thermohaline circulation may not be of direct importance, and the circulation within the basin may actually be driven by small frictional stresses exerted upon it by the thermohaline circulation above sill depth, in much the same manner as in an ordinary wind-driven circulation.

Finally, the vertical distribution of vertical velocity must be of great importance in determining the vertical temperature and salinity structure

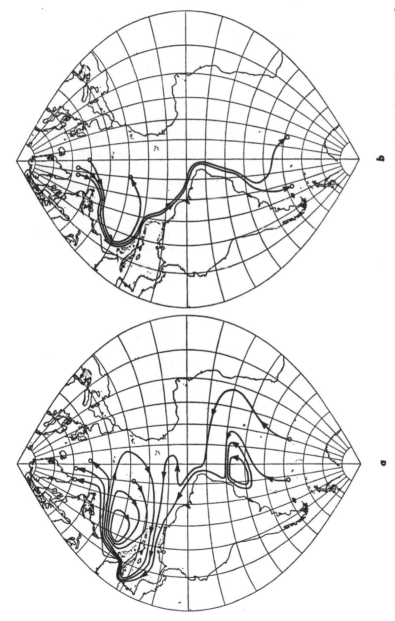

Fig. 82. Schematic charts of the transport in the upper layers (a) and lower layers (b), as inferred from data presented in fig. 79 and from the hypotheses discussed in the text. No attempt has been made to preserve the details of current structure; the figure is meant as a rough first approximation to serve as an intermediate link between the completely idealized models, such as that indicated in fig. 78, and the more complex detail of the actual ocean, as partially indicated in fig. 79.

in the ocean. Were the eddy diffusion coefficients independent of depth, the depth of maximum w might nearly correspond to the bottom of the main thermocline.

In summary, then, the line of argument and interpretation of data presented here, although by no means conclusive, do suggest: (i) that the interior of the subtropical and equatorial Atlantic Ocean is essentially wind-driven, from a dynamical point of view; (ii) that in subpolar regions thermohaline processes are important; (iii) that in the western boundary currents, effects due to the mass-conservation requirements imposed by both the wind stress and the thermohaline process are important, even in subtropical and equatorial latitudes; (iv) that the portion of the surface-layer western current which is driven by wind stress is intimately connected with Munk's gyres, or a cell-like structure of wind currents; (v) that the thermohaline portion of the surface-layer western current flows from 40° S. latitude to 40° N. latitude almost unchanged in transport; (vi) that the deep circulation at all latitudes is essentially thermohaline, but in subtropical and equatorial latitudes is confined to a narrow western current of the (perhaps inertial) boundary-layer type; and (vii) that, therefore, the mean vertical flux of mass in the interior of subtropical regions at depths of 1000–2000 m. is probably considerably less than that used in the theoretical rectangular ocean (less than 10 cm./day).[4]

[4] Since the preparation of the manuscript for this book, I have written and published a review article which supplements the ideas presented in this chapter: 'A Survey of Ocean Current Theory', *Deep-Sea Research*, 4 (1957):149–184. Also, I have attempted to make an analysis of hydrographic data in the North Atlantic according to equation (27) under the title: 'On the Determination of the Depth of No Meridional Motion', *Deep-Sea Research*, 3 (1956):273–278.

Chapter Twelve

GENERAL REMARKS

The central North Atlantic Ocean is covered to a depth of about 700 m. with a layer of warm water, slowly drifting toward the southwest under the combined influence of the stress of the wind and the earth's rotation. This central body of warm water, called the Sargasso Sea, does not reach the western coasts of the Atlantic Ocean (the eastern coast of North America), but is bounded some distance offshore by the Gulf Stream—a narrow, intense, northeastward-flowing current which returns to the north again the southward-driven Sargasso Sea water that has passed through the Caribbean and has turned through the Florida Straits.

The Gulf Stream flows along the western boundary of the warm Sargasso Sea surface water. As the Stream turns toward the east, off the Grand Banks, it acts as a kind of dynamic barrier, or dam, which, by virtue of Coriolis forces, restrains the warm Sargasso water from overflowing the colder northern water of the North Atlantic. The water in the Stream is not significantly different in temperature from the large mass of warm water which lies to the right of its direction of motion. The Gulf Stream is not an ocean river of hot water. The intensity of flow of the Stream, the Stream's direction, and its temperature are not primary climatic factors in determining the climate of Europe; but the role which it plays in determining the northern boundaries and average temperature structure of the Sargasso Sea must be of critical climatic importance. However, we do not yet know how this role is physically connected with such observables as total transport.

Certain remarks concerning the very nature of oceanographic research itself are made in a later section of this chapter.

JUSTIFICATION FOR TREATING THE GULF
STREAM SYSTEM AS AN ENTITY

Perhaps it has occurred to the reader to wonder whether there is any justification for treating the Gulf Stream System as a physical system in any sense separable from the general circulation of the North Atlantic Ocean. If the Gulf Stream were a river of very warm water flowing through a colder environment, there would be little doubt that it should be considered a distinct physical phenomenon. As we have seen in the preceding chapters, the most clear-cut feature of the entire system is that it is really a flow along the very edge of the juncture of a mass of cold water and a mass of warm water. Therefore, one is led to inquire whether the primary physical phenomena to discuss are not the origin and nature of the two contrasting water masses, and whether it is not proper to regard the Gulf Stream current system as merely a secondary feature associated with their zone of contact.

From a historical point of view, the Gulf Stream has always been the primary feature of interest in the western North Atlantic, and the efforts of the first scientific oceanographic explorations were concentrated on it because it was so clear-cut an entity. 'It' could be crossed in a day; 'its' total water transport could be determined; 'its' changes in position could be plotted; 'it' had an edge, a core of high velocity, a countercurrent, and so on. Compared to other regions of the ocean, it seemed one where techniques of measurement, such as the indirect velocity determinations based on the geostrophic approximation, could be made with a reasonable degree of precision. When one considers the vast central oceanic areas, or the eastern sides of oceans, where the current features are very faint and indistinct—almost beyond the reach of standard means of measurement—it is no wonder that oceanographers on the western sides of oceans often consider themselves especially fortunate to possess such strong current systems near their coasts and, as a result, have spent much more time cruising in these strong currents than in obtaining detailed information about the great water masses on either side.

Also, from the point of view of theoretical studies, the Gulf Stream System has a distinct reality of its own. The outstanding feature of these theories (discussed in Chapters VII and VIII) is that the transport solutions are independent of any assumptions regarding the density structure. The Gulf Stream would exist, according to these theories, whether it is driven by wind stress or by a thermohaline process. It would exist whether the ocean were half as deep as it is, or ten times as deep. The velocities would be different, but the pattern of the transport lines would be the same. The Gulf Stream System emerges from these theories as a boundary phenomenon occurring at a certain boundary of the oceans where high-order derivatives in the governing equation (elsewhere negligible) are locally important. In this sense, the Gulf Stream 'exists' as a physical entity in the same sense as the boundary layer next to an aerofoil exists: it is a recognizable part of a larger physical system.

The idea of a western boundary current—whether it be viscous or inertial—is of basic importance because it simplifies the physics and mathematics of the central regions of the ocean. The higher-order terms in the vorticity equations are restricted to a narrow coastal region, and the remainder of the ocean can be studied with very simple equations (as we have already seen in Chapter XI). This possibility of emancipating the interior solution from boundary conditions at the west may very well be the most important lesson to be learned from the study of the Gulf Stream.

CLIMATE AND THE GULF STREAM

There is scarcely any more firmly rooted idea in the mind of the layman than the notion that the Gulf Stream keeps the European climate warm. So long as it was believed that the Gulf Stream was a river of warm water, this idea did make sense. It is no longer possible to be so certain of the direct climatological influence of the Gulf Stream, for it now seems that it is not so much the Stream itself that is important, as the position and temperature of the large mass of warm water on its right-hand flank.

Any argument which purports to relate climate to the transport of the Gulf Stream must include some account of the physical relations between the Stream and the density (and thermal) structure of the water on either side. In fact, Iselin (1940) went so far as to speculate that warming of the European climate might actually be least during periods of increasing transport of the Gulf Stream. This idea was based on the hypothesis that the processes which produce the warm surface masses of the Central Atlantic Water are more or less constant in time, and hence, owing to the geostrophic relation, an increasing transport of the North Atlantic Gyre must be accompanied by simultaneous deepening of the thermocline in the Sargasso Sea and a radial shrinking of the current system. In this way, warm surface water would be withdrawn from the north, and the European climate might be colder. Conversely, a weakening of the transport of the current system would, Iselin reasoned, be accompanied by a rising thermocline throughout the Sargasso Sea, and the excess warm water would force the current system radially outward and farther northward, and might even succeed in discharging quantities of surface water to high latitudes, thus warming the European climate. There is no convincing evidence to prove that this (or any other) sequence of events actually takes place, nor can we even discuss the hypotheses very satisfactorily; in particular, we know very little about the *rates of formation of water masses*. But these arguments are of special interest, because they confound the popular idea of the role of the Gulf Stream in European climate. For all we know, the European climate might actually be warmer if the direction of rotation of the North Atlantic Eddy were reversed.

During the past thirty years there has been an increase of about 2° in the surface temperatures of the Norwegian Sea, and also a decrease of perhaps 50 m. in the depth of the 10° C. isotherm throughout the Sargasso Sea. (The statistical significance of this change of position of the isotherm is doubtful.) These two isolated bits of information suggest that the North Atlantic surface circulation is slowing down and that the Gulf Stream transport is consequently decreasing, and that this, in turn, results in greater warming of the coast of Europe. This indication favors Iselin's proposal.

The theory of wind-driven ocean currents suggests that the effect of diminished winds would make the North Atlantic circulation shallower in the middle, but would not change its radius. Toward the periphery of the circulation the isotherms might become deeper. However, it is important to remember that the theory, in its present state of development, deals only with integrated velocities, and hence that there is no clear theoretical indication of the ways in which temperature, heat transport, and surface-current velocity might vary with the wind. Moreover, the climate and winds over the ocean are not unaffected by the state of ocean currents. As a result, the theory of wind-driven currents permits us to draw inferences about the change of currents resulting from only very slight changes of the winds. It seems to me quite reckless to assert, for example, that some current of the Carboniferous Glacial Period, flowing in an ocean which does not even exist now, had a greater or lesser transport than the present-day Gulf Stream. The more these hypothetical oceans of the past depart from the configuration and climate of the oceans of today, the less we can hope to explain and describe their behavior.[1] When the disparity between supposed conditions of the past and observed conditions of the present becomes as great as that envisaged in Chamberlain's theory of the reversal of the deep circulation due to the sinking of highly saline water at the equator, the unhappy physical oceanographer can make no constructive comment.

---

[1] Many examples of purely verbal theories that invoke physical processes with which our present primitive theoretical models cannot yet cope are given in Chapter III of C. E. P. Brooks's book *Climate Through the Ages* (1949). Another example, published since this text was written, is 'A Theory of Ice Ages', by M. Ewing and W. L. Donn, *Science*, 123 (1956):1061–1066.

SOME REMARKS OF A POLEMICAL NATURE

By far the greatest part of physical oceanographic knowledge was accumulated by the following process:

First, a grand cruise, or expedition, brings back data obtained at many hydrographic stations.

Secondly, extensive plots, graphs, and tabulations of the data are made and published for the benefit of future generations (e.g., in *Bulletin Hydrographique*).

Thirdly, certain of the more striking features of the data plots are noted.

Fourthly, some plausible hypotheses are advanced to explain them.

At this stage the procedure had usually exhausted the energy of those involved, and almost always the funds, and the study usually stopped. In a few projects, such as the *Discovery* Antarctic studies, the Scripps Marine Life Program, and the Woods Hole Gulf Stream studies, attempts were made repeatedly (third step) to bring the striking features into better focus, the first and second steps being repeated in various combinations and permutations, with the goal of presenting the main features of the findings in great detail. Thus, much of the effort in the Gulf Stream work has been devoted to attempts to detect changes in the total transport of the Stream, to surveys of the shapes of meanders, and to determinations of the 'width' of the Stream and of the transverse surface velocity profile. The plausible hypotheses (fourth step) are modified as needed, to avoid conflict with the observations. Sometimes the choice of hypothesis is easy and the modification slight, because a variety of different plausible hypotheses all fit the observations.

These first four steps are absolutely necessary, of course; without them we would know nothing at all about the ocean as it really is. But for the full development of the science there must be one additional step:

Fifthly, the plausible hypotheses must be tested by specially designed observations; in this way theories can be rejected or accepted, or may be modified to become acceptable.

As an example of an untested hypothesis one may take any particular choice of the 'depth of no motion' for dynamic computations of the Gulf Stream (see Chapter XI). A knowledge of the truth or falsity of this hypothesis will become indispensable once the theoretical models begin to approach reality. To make a satisfactory test will exercise our ingenuity. The surface currents are too strong to permit the making of accurate direct current measurements from an anchored ship. Oceanographers may have to devise and employ some new observational technique in order to make this important test, but the great obstacles should not blind us to the necessity of undertaking it. Perhaps the most promising technique for direct deep-current measurements underneath the Gulf Stream is a method which is now being refined by Dr J. C. Swallow at the British National Institute of Oceanography and was recently described at a London conference on the deep ocean circulation. A small float, so ballasted as to reach and remain at a fixed predetermined depth, carries a battery-powered ultrasonic noisemaker. A ship with a directional hydrophone array can track the drift of the float, and hence obtain direct measures of the drift of deep currents. Whether this technique can be refined to measure currents as accurately (say to 0·5 cm./sec.) as will be needed to determine the level of no motion below the Gulf Stream, remains to be seen; but it is important that this test should be made.[2]

Another example of an untested hypothesis is the entire conception that large-scale lateral mixing occurs in the ocean. As yet, there does not seem to be àny obvious way to test this idea.

I should like to make it clear, finally, that I am not belittling the survey type of oceanography, nor even purely theoretical speculation. I am pleading that more attention be given to a difficult middle ground: the testing of hypotheses. I have not explored this middle ground very thoroughly, and the few examples given in this book may not even be the important ones; but perhaps they are illustrative of the point of view in which attention is directed not toward a purely descriptive art, nor toward analytical refinements of idealized oceans, but toward an understanding of the physical processes which control the hydrodynamics of oceanic circulation. Too much of the theory of oceanography has depended upon purely hypothetical physical processes. Many of the hypotheses suggested have a peculiar dreamlike quality, and it behooves us to submit them to especial scrutiny and to test them by observation.

[2] For a report of very recent work of this kind see Chapter XI, footnote 2.

Chapter Thirteen

RECENT DEVELOPMENTS

Since the first edition of this book was published, much new work has been done on the Gulf Stream. The purpose of this new chapter is to review this new work and bring it into focus. First we will discuss the new observational material. The Multiple Ship Survey of 1960 was different in concept, design, and results from the earlier survey of 1950. It has been reported upon by Fuglister (1963). It is an extension of the subject of Chapter V.

Next we consider the various attempts to map out the deep current structure under and near the Gulf Stream. In addition to the short note of Swallow and Worthington (1957) that is mentioned on page 163, a more detailed treatment of the results has now been published by these same authors (1961). We also have newer measurements made by Volkmann (1962), those made by Woods Hole staff members on the Multiple Ship Survey of 1960 and reported by Fuglister (1963), and most recently measurements near Cape Hatteras by J. Barrett (in press).

New ideas about fluctuations in the Stream and about the mechanism of meanders have been proposed by Webster (1961) and Warren (1963). Webster claims to have found evidence that eddy processes south of Cape Hatteras play a role which is opposite to the dissipative one normally associated with turbulence in a stream. Warren claims to have shown that meanders east of Cape Hatteras are tightly coupled with bottom topography through the conservation of vorticity. Both of these ideas may have profound significance for our picture of the physical processes which govern the Stream.

Considerable interest has been shown in the theory of western boundary currents in the past few years. In particular, inertial boundary currents—the subject of Chapter VIII—have become quite a popular subject for theoretical study. These have centered about the difficulty of bringing lines of constant transport function out from a western inertial boundary layer into the ocean interior—a difficulty which was first described by Morgan (1956). In trying to circumvent this difficulty Carrier and Robinson (1962) were led to propose a rather unoceanic solution which they claim avoids the main difficulties encountered when trying to fit together inertial boundary layers and geostrophic interiors to form closed solutions for completely bounded basins. This flow-pattern is so unacceptable that it has stimulated further analytical studies by Greenspan (1962), Moore (1963), and Ilyin and Kamenkovich (1963), as well as numerical studies by Bryan (1963) and Veronis (1963b).

Chapter XI could be entirely rewritten, as so much has happened during the past few years to clarify the theory of the thermohaline circulation in the open ocean. Unfortunately this work impinges only indirectly upon the Gulf Stream, and therefore it is not referred to in detail, but only briefly reviewed.

The recommendations for future work which were made in Chapter XII are expanded and reformulated in the light of the growing capabilities of buoyed instrument technology and of our growing recognition of the importance of large-scale turbulent processes in the dynamics of the ocean circulation. As the reader will see, the diagnosis and prescriptions are changing with time.

'GULF STREAM '60'

During the spring of 1960 oceanographers of the Woods Hole Oceanographic Institution and the International Ice Patrol coöperated in a large-scale survey of the Gulf Stream east of Cape Hatteras which was similar in scope, though not in detail, to the Multiple Ship Survey called 'Operation Cabot' in 1950 (pp. 51 ff.). The name which somehow became attached to this survey was 'Gulf Stream '60,' and the first comprehensive report that has appeared in print was prepared by Fuglister (1963). This valuable paper includes vertical profiles of temperature, salinity, and oxygen, and tabulations of the original station data, as well as many special charts. Although it is quite impossible to review the paper in detail in this chapter, we can outline some of the highlights.

During the period April–June, 1960, four ships obtained hydrographic station data from the surface to the bottom at the points indicated in fig. 83. Most of the stations were occupied during April, but enough were

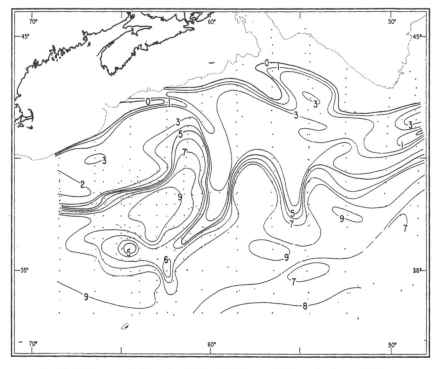

Fig. 83. 10° isotherm depth, meters × 100, 'Gulf Stream '60'. From Fuglister (1963).

occupied during the following two months to make it clear that the large meander centered on the 60° W. meridian was nearly stationary. This meander is much larger than anything observed before (see fig. 28 for comparison with others in the past), especially if one follows the 700 or 800 m. contour of the 10° isotherm depicted in fig. 83. Samples of the vertical profiles of temperature are shown in figs. 84 and 85. They run southward from the coast. Fig. 84, section I, is the westernmost line of stations. Fig. 85, section III, is the third line of stations counted from the west; it crosses the little cold eddy shown at the 'toe' of the large meander in fig. 83. There are many details revealed in this survey which merit discussion but which cannot be covered here. For example, Fuglister points out that, although section I is the simplest profile of this series, even here the Gulf Stream limits cannot be precisely drawn.

Stations 5880, 81 and 82, for instance, are definitely in the boundary zone, which appears to reach from the surface to the bottom, but should the zone be extended to stations 5878 and 5883 on the basis of the continued slope of the isotherms in the water beneath the main thermocline? Also, should the relatively slight disturbance in the thermocline around station 5885 be considered part of the boundary zone?

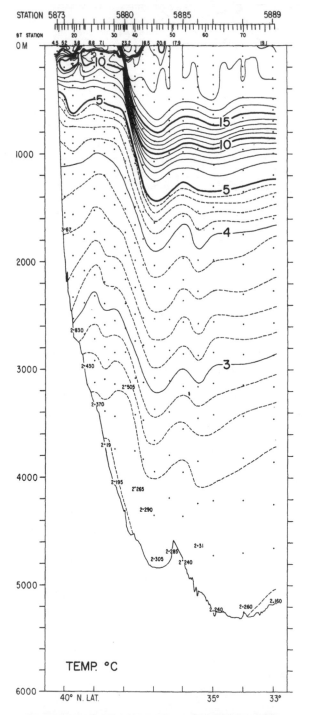

Fig. 84. Section I, 68° 30′ W. longitude. From Fuglister (1963).

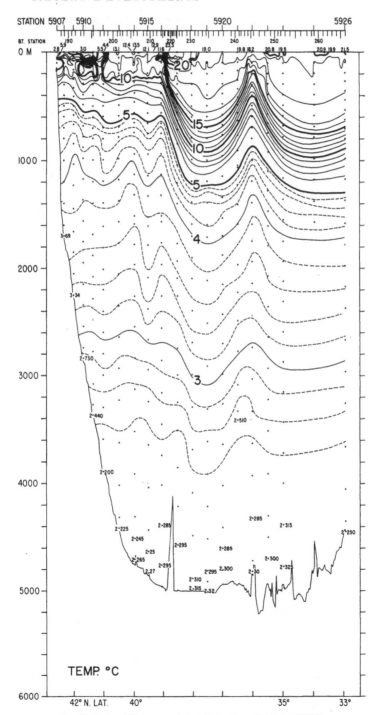

Fig. 85. Section III, 64° 30′ W. longitude. From Fuglister (1963).

Or, again, concerning the warm core which is often regarded as a fixed feature of the Stream:

> The temperature profile shows clearly the surface warm core of the Gulf Stream but what is the significance of the smaller core of warm water north of this disturbance? On section III, the 'disturbance' is more pronounced and located farther to the south; thus this section crosses the boundary zone, as defined by the sloping thermocline, in three places; on the other hand, only one well-defined warm core appears on the section.

Figs. 86 and 87 show salinity and oxygen at the section III. Fig. 88 shows volume transport across section III using the bottom as reference level. The stream appears to be double because of the eddy which is crossed south of the Stream. Fuglister gives a summary of the transport calculations given in table 9. All these values are based on data from the first phase of the study.

TABLE 9
VOLUME TRANSPORTS 10^6 M³./SEC.: 'GULF STREAM '60'

Section	1	2	3	4	5	6	7	8	9	C.G.[a]
Between latitudes	38° 20' 37° 00'	39° 01' 37° 00'	39° 02' 37° 30'	39° 34' 38° 00'	42° 20' 39° 28'	41° 31' 38° 30'	41° 01' 38° 32'	41° 00' 39° 02'	40° 00' 37° 28'	
Bottom 2000 m.	137 89	106 66	88 64	76 57	50 33	80 53	77 55	52 37	82 58	51
Slope water Current latitudes						43° 19' 42° 00'	42° 59' 41° 29'	44° 00' 42° 55'	42° 20' 41° 30'	
To 2000 m.						7	8	2	9	4

[a]C.G. values from SOULE et al. (1961).

A measure of the increase in transport that would be obtained if the observed deep-current velocities were used in the computations indicates that the transport of the Gulf Stream on section III would change from 88 to 147 × 10^6 m.³/sec. If the Gulf Stream transports these huge amounts of water, then, in order to satisfy continuity, there must also exist deep-water movements of considerable magnitude elsewhere in the System which we have not mapped out, and which do not find any simple explanation in terms of our theoretical studies to date.

One of the very striking features of these profiles, as well as of the others which we do not reproduce here, is the fact that the strong horizontal temperature gradient occurs under the Gulf Stream at all levels, even to the bottom. Therefore, we expect a strong current shear at depth, and this was measured by neutrally buoyant floats during Gulf Stream '60 and was found to flow with the surface current all the way to the bottom. However, it should also be noticed that in these sections there is a small region of minimum temperature along the edge of the Continental Rise (2.19°C. on section I, 2.20°C. on section III, for example)

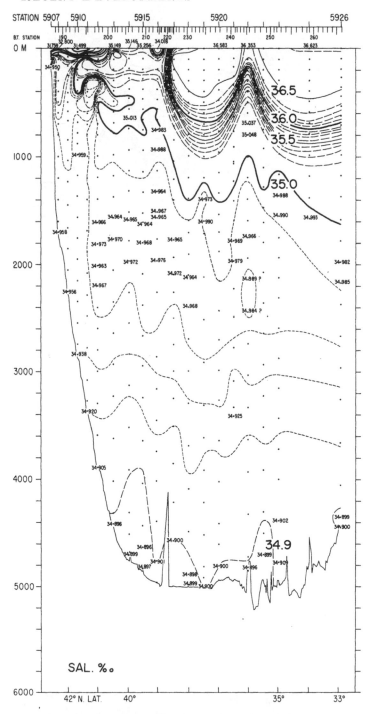

Fig. 86. Section III, 64° 30′ W. longitude. From Fuglister (1963).

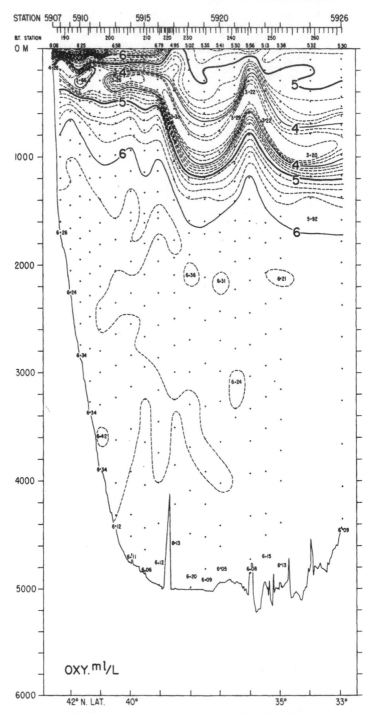

Fig. 87. Section III, 64° 30′ W. longitude. From Fuglister (1963).

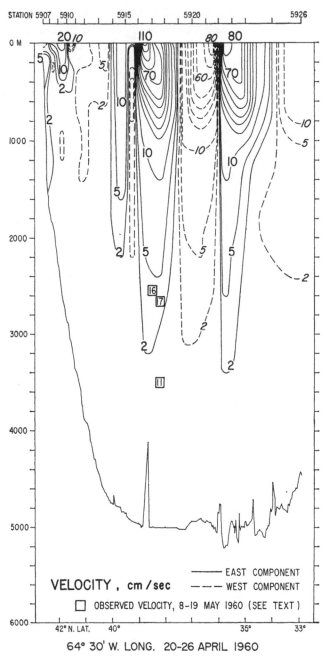

Fig. 88. Velocity profile of section III. From Fuglister (1963)

and this does not lie directly under the main horizontal temperature gradient of the Stream, and therefore appears to be something else entirely. As we will see, we think of this as the water flowing south and westward to feed the undercurrent off the Blake Plateau.

One of the interesting features of the Gulf Stream is the mass of fairly homogeneous water which overlies the thermocline on the right-hand side of the Gulf Stream. It is remarkably stable and constant in temperature and salinity over a large area and over the years despite the fact it is in contact with varying winter conditions at the surface each year when it is partly renewed. The remarkable constancy of this water mass has been the subject of papers by Worthington (1959) and by Schroeder et al. (1959).

An important work to which anyone interested in the Gulf Stream and Atlantic Ocean will frequently want to make reference is a handsome colored atlas of the profiles of temperature and salinity obtained during the I.G.Y. and prepared by Fuglister (1960).

UNDERCURRENT OBSERVATIONS

The set of measurements on the deep currents under the Gulf Stream at the edge of the Blake Plateau, made by Swallow and Worthington using the *Discovery II* and the *Atlantis* in the spring of 1957, and briefly mentioned in the footnote on page 163 of this work, has since been published in further detail (Swallow and Worthington, 1961). The method of measurement using the neutrally buoyant pingers is described, and there is also a careful description of each of the accompanying hydrographic

TABLE 10
SUMMARY OF CURRENT MEASUREMENTS WITH NEUTRALLY BUOYANT FLOATS, MARCH–APRIL, 1957

Float	Launched (time and date)	Last fix (time and date)	Intended depth (m.)	Mean observed depth (m. ± st. dev.)	Mean velocity (cm./sec. ± st. dev.)	Direction (°T. ± st. dev.)	Remarks
B	1000/6	0743/11	2000	2040 ± 70	0.33 ± 0.11	108 °T ± 18°	
D	1715/17	0718/22	2500	2550 ± 40ᵃ	4.27 ± 0.21 1.88 ± 0.11	201 °T ± 1.4° 235 °T ± 4°	2206/17–0320/19 0636/19–0718/22
E	1527/20	1717/22	1500	1480 ± 50	6.42 ± 0.47 6.50 ± 0.29	308 °T ± 4.7° 231 °T ± 2.3°	1825/20–2135/21 0010/22–1717/22
F	0954/23	0733/25	2500	2620 ± 80ᵃ	8.99 ± 0.53	190 °T ± 2.3°	
G	2050/23	1032/26	2500	2600 ± 50ᵃ	4.41 ± 0.26 7.08 ± 0.44	218 °T ± 3.3° 203 °T ± 2.3°	0002/24–0314/25 1110/25–1032/26
H	0815/26	0605/29	2800	2910 ± 70ᵇ	18.36 ± 0.28	182 °T ± 1.0°	
I	0854/26	1040/29	2800	2760 ± 190ᵇ	6.02 ± 0.27 12.62 ± 0.39	216 °T ± 2.6° 196 °T ± 1.3°	1424/26–0555/27 1715/27–1040/29
J	1110/30	1103/2	2800	2900 ± 120ᵇ	10.24 ± 0.18 9.42 ± 0.22	204 °T ± 1.7° 185 °T ± 11°	1347/30–0034/1 0858/1–1103/2
K	1156/30	0935/31	2800	2770 ± 200ᵇ	12.95 ± 0.59	207 °T ± 5.2°	

ᵃMean for D, F, and G = 2580 m.
ᵇMean for H, I, J, and K = 2840 m.

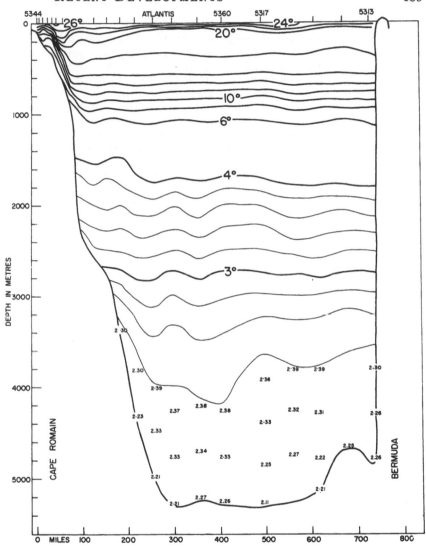

Fig. 89. Temperature section, Cape Romain to Bermuda, made by the *Atlantis* in June, 1955. From Swallow and Worthington (1961).

sections made by the *Atlantis*. Even as early as June, 1955 (see fig. 89), it was evident that a strong current lies at the foot of the shelf off Blake Plateau, as shown by the strong slope of the deep isotherms. Moreover, this slope is displaced from the surface currents so that a ship above the deep current could be maneuvered fairly easily. The results of the measurements are summarized in table 10.

Concerning this table, Swallow writes:

> The seven measurements at depths below 2000 m. show southward movement, and although large variations were found at given depths, a general increase towards the bottom is indicated. It seems likely that the mean depths, of 2580 m. for floats D, F and G, and 2840 m. for H, I, J, and K are more reliable measures of the depths reached by these floats than the individual determinations are. The large scatter, from 2760 m. to 2910 m. at a nominal depth of 2800 m., is probably not genuine, in view of the close similarity of design of the floats and the accuracy of adjustment of densities. The very small mean current observed with float B, shows that, at that time and place, the level of no motion was very close to 2000 m. in depth.

Fig. 90 shows one of the *Atlantis* sections (the fourth) as an example.

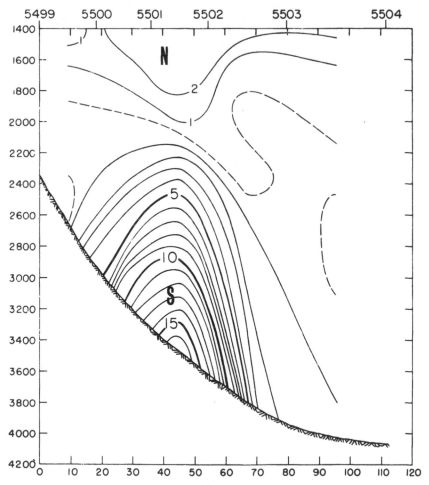

Fig. 90. Section IV. From Swallow and Worthington (1961).

The contours are lines of constant geostrophic velocity computed with a reference level referred to float D, which is also indicated on the section. The southerly component measured by the float was 3.7 cm./sec. at 2580 m., the reference level at 1950 m. The volume transport of the southerly current was 4.1×10^6 m.³/sec.$^{-1}$.

Following the measurements off Cape Romain in 1957, Volkmann (1962) made more measurements farther downstream, south of Cape Cod in 1959 and 1960, within regions of steeply sloping deep isotherms, a common phenomenon along this coast. Volkmann used the measurements to establish the reference level for the computation of geostrophic velocities and extended these references across the sections for the computation of transports. There is a large difference in the transport computed for the two years, and both are appreciably greater than that reported by Swallow and Worthington off Cape Romain.

The areas of measurement are shown in fig. 91, and the results obtained are shown in table 11. The region is clearly rather complex, and it is uncertain whether there is a permanent west and south flowing undercurrent revealed by these measurements.

TABLE 11
VOLKMANN'S MEASUREMENTS USING NEUTRALLY BUOYANT FLOATS

Designation	Latitude	Depth (m.)	Date	Hours tracked	Distance followed (km.)	Velocity cm./sec.	Velocity toward	East-west component (cm./sec.)
Float 8	37° 30'	1945 ± 595	19–21–VII–1959	41.6	32.2	21.5 ± 1.0	248°	19.9
4	37° 30'	3200 ± 980	25–26	28	18.6	18.4 ± 1.8	230°	14.1
9	38° 50'	1900 ± 535	28–29	22.2	9.1	11.4 ± 1.1	263–213°	8.0
Float I	36° 50'	1330 ± 540	17–19, VII–1960	36	30.5	23.5 ± 0.3	113°	21.6
II	36° 47'	2160 ± 945	18–19	23.3	16.1	19.2 ± 0.8	106°	18.4
III	38° 54'	1850 ± 610	20–22	52.5	22.3	11.8 ± 0.5	294°	10.8
IV	39° 18'	1910 ± 680	22–26	92.5	24.6	7.4 ± 0.5	326–089°	—
V	38° 00'	2120 ± 640	23–24	23.3	8.4	10.0 ± 0.4	130°	7.7
VI	38° 00'	2460 ± 585	23–24	20.3	4.8	6.5 ± 0.8	125°	5.3

In addition measurements were made during the Gulf Stream '60 survey by the *Atlantis* and *Crawford*, as reported by Fuglister (1963), and are shown in table 12.

The tracks of the deep *Atlantis* floats are shown in fig. 91. As Fuglister says,

In spite of the uncertainties of the depth calculations, there is no question but that the floats were at depths well below 2000 m., that they were in the Gulf Stream and that over a period of 11 days the deep flow was essentially in the same direction as the flow at the surface and at a depth of 700 m. The first float, at a calculated depth of 2650 m., was tracked for 116 hr. at an average speed of 17 cm./sec. The second float, at 3500 m., moved at 11 cm./sec. for 42 hr. The third float, at a calculated depth of 2550 m., was the most interesting: it was followed for 83 hr. at an average speed of 16 cm./sec.; it headed toward Kelvin Sea Mount and then curved around to the north, obviously deflected by this obstacle. A

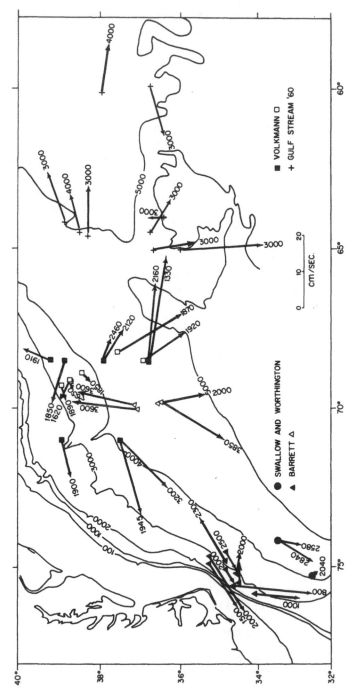

Fig. 91. A chart of all the deep pinger tracking results to date.

segment of the 3000 m. depth contour of this sea mount is shown in [fig. 92] for comparison with the float track. All these direct current measurements of the *Atlantis* and *Crawford* showed the subsurface currents in the Gulf Stream to be essentially in the same direction as the surface flow.

Therefore we are faced with quite a different state of affairs east of Hatteras than south of Hatteras. The situation is complicated by the fact that measurements of deep currents by Swallow and Crease off Bermuda in 1959 and 1960 showed rather large velocities even far away from the Gulf Stream in regions where we might naturally suppose that the amplitude of the deep currents would be much smaller. These studies have not yet been published, but amplitudes of 20 cm./sec. at depths of 4000 m. off Bermuda were not uncommon. This kind of amplitude is disconcerting: it suggests that perhaps the similar velocities

TABLE 12
DIRECT CURRENT OBSERVATIONS: 'GULF STREAM '60'

Number	Intended depth	Cal. depth	Date		Elapsed hours	Position		Distance (miles)	Direction	Speed (cm./sec.)
			First fix	Last fix		First fix	Last fix			
Atlantis										
1	3000	2650	0745 8 May	0400 13 May	116.2	38° 21′ N. 65° 11′ W.	38° 21′ N. 64° 22′ W.	39	090°	17.2
2ª	4000	3500	1540 11 May	1000 13 May	42.3	38° 25′ N. 64° 34′ W.	38° 30′ N. 64° 25′ W.	9	058°	11.0
3b	3000	2550	1615 15 May	0330 19 May	83.2	38° 41′ N. 64° 20′ W.	38° 56′ N. 63° 55′ W.	26	053°	16.1
4	4000	3580	0600 3 June	0640 6 June	72.0	37° 57′ N. 61° 03′ W.	37° 52′ N. 60° 32′ W.	25	102°	17.7
5c	3000	—	0950 7 June	1300 7 June	3.2	37° 42′ N. 60° 29′ W.	37° 39′ N. 60° 26′ W.	—	—	—
6	3000	—	2045 9 June	1530 11 June	42.8	36° 44′ N. 59° 46′ W.	36° 42′ N. 59° 57′ W.	10	255°	12.0
Crawford										
7d	700	—	0920 5 May	0118 8 May	64.0	37° 49′ N. 68° 22′ W.	38° 14′ N. 66° 23′ W.	105	100° 035°	105.0 60.0
8e	400	—	0910 10 May	0911 12 May	48.0	38° 05′ N. 68° 24′ W.	37° 49′ N. 67° 28′ W.	47	090° 115°	51.0 51.0
9f	700	—	1300 13 May	1100 17 May	92.0	38° 41′ N. 63° 22′ W.	39° 15′ N. 61° 30′ W.	95	085° 060°	90.0 45.0
10	3000	2480	1748 2 June	1025 10 June	184.5	37° 15′ N. 65° 01′ W.	38° 56′ N. 64° 46′ W.	35	160°	10.0
11g	3000	4530	1015 4 June	1400 10 June	147.8	36° 46′ N. 64° 28′ W.	36° 42′ N. —	—	—	—
12	3000	2160	0615 7 June	1655 7 June	10.7	36° 44′ N. 64° 37′ W.	36° 42′ N. 64° 35′ W.	3	120°	14.0
13	3000	—	0550 9 June	1740 10 June	35.8	36° 32′ N. 64° 08′ W.	36° 35′ N. 64° 08′ W.	3	360°	4.0
14	3000	—	1500 11 June	2015 12 June	29.2	36° 03′ N. 65° 05′ W.	35° 52′ N. 65° 04′ W.	11.5	175°	20.0

For the longer runs, 7, 8, and 9, the mean direction and speed during both the first and last parts of the runs are shown.

ª Slight cyclonic curvature.

b Anticyclonic curvature (radius 10 miles) around northwest side of Kelvin Sea Mount. Velocity increased to about 20 cm./sec. while near sea mount.

c Too short a time for estimate of current.

d Rapid speed decrease after 48 hr. cyclonic curvature.

e Slight anticyclonic curvature.

f Gradually decreasing speed with cyclonic curvature.

g Slight random movements recorded, but this float was probably grounded.

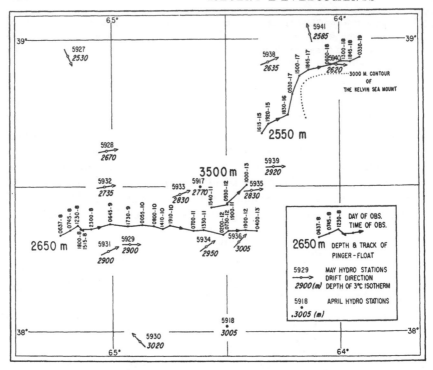

Fig. 92. *Atlantis* track of pinger floats and station positions, May, 1960. From Fuglister (1963).

measured off Cape Romain in 1957 might be transient. Against this, however, is the minimum temperature of water observed in the south-ward-moving current there; a phenomenon which does not have a counterpart in the eddies observed off Bermuda. The high amplitude transient motions off Bermuda, furthermore, do not seem to be present in the eastern basin of the Atlantic, and hence one is led to wonder if they are associated in some way with the Gulf Stream, perhaps as the last vestiges of eddies thrown off months or years previously.

In an effort to clarify the problem of the patterns of deep flow beneath and near the Gulf Stream, J. R. Barrett in the fall of 1962 tracked thirteen floats at sections near Cape Hatteras. Although the results are not yet published they were presented at a lecture at Woods Hole (see table 13), and they confirm the existence of a strong deep southward flow similar to that reported by Swallow and Worthington in 1957. Barrett has also made a careful study of the water-mass properties of the water under the Gulf Stream and up to the continental shelf and is able to show that the southward-flowing current is noticeably higher in

TABLE 13

SUMMARY OF BARRETT'S DIRECT CURRENT OBSERVATIONS USING NEUTRALLY BUOYANT FLOATS

Float	Nominal depth (m.)	Launch date	Launch position		Tracking time (hr.)	Velocity		Remarks
			Lat.	Long.		cm./sec.	Direction	
1	1200	4 × 62	33° 58' N.	75° 49' W.	4.0	19.3	230°	Too short a time
2	1000	5 × 62	34° 04'	75° 46'	22.0	13.1	190°	Depth by bottom refl. = 1045 m.
3	800	6 × 62	34° 29'	75° 38'	38.0	21.1	205°–180°	Slight anticlockwise curve
4	2000	6 × 62	34° 23'	75° 29'	49.0	5.5	065°	
5	750	8 × 62	34° 44'	75° 27'	—	—	—	Lost after one fix
6	2000	8 × 62	34° 39'	75° 18'	53.0	12.0	220°	
7	2000	9 × 62	34° 32'	75° 06'	32.0	5.5	090°	
8	2000	11 × 62	34° 56'	75° 02'	78.0	6.7	220°	
9	2300	12 × 62	34° 54'	74° 54'	48.0	10.1	055°	
10	1500	22 × 62	35° 14'	74° 50'	31.5	15.2	245°–205°	Anticlockwise curvature
11	2000	22 × 62	35° 10'	74° 43'	27.0	17.7	220°–248°	Converging with #10
12	2500	23 × 62	35° 03'	74° 31'	3.2	20.0	185°	Too short a time
13	2500	24 × 62	34° 58'	74° 34'	28.0	6.2	235°	

oxygen, lower in salinity, and lower in temperature than surrounding water masses, that it tends to hug the place where the continental escarpment joins the continental rise, rather than flowing directly bebeneath the Stream. It will be interesting to observe this subject during the next few years as successive complications and revelations unfold.

MEANDERS AND BOTTOM TOPOGRAPHY

One of the supporting pieces of evidence that the currents below the strong parts of the Gulf Stream east of Cape Hatteras are in the same direction as the surface flow is the work of Warren (1963). The effect of varying depth is similar to that of varying Coriolis parameter—as has been known by tidal theorists for a long time—but little use has been made of this fact by oceanographers. As a matter of fact, most ocean currents do not seem to be very grossly influenced by bottom topography; even such pronounced features as the Mid-Atlantic Ridge do not deflect the Equatorial Currents which cross them. However, where a current does reach all the way to the bottom it seems certain that the effect must be noticeable. After leaving the shelf at Cape Hatteras the Stream impinges at an angle upon the slope off the continental shelf, and Warren believes that it coasts up and down this slope in a standing meander pattern. Depending upon the initial angle of incidence, different patterns are possible, and Warren has invented an approximate numerical technique for computing these paths over the known bottom topography. The results of the calculations, which Warren has compared to all available observations of the meander pattern, agree very well with observed paths of the Stream. Warren regards this as reasonable proof that Gulf Stream meanders east of Hatteras are not a result of instabilities, that the flow is in the same direction at all depths, and that in a very real sense the Stream does not 'leave the coast' after passing Hatteras, as many writers generally interpret the situation there to imply. The

Stream seems to 'feel' the bottom at least as far east as the longitude of the Grand Banks of Newfoundland. The hydrodynamical situation is difficult to model, considering the degree of natural complexity which the calculation is supposed to predict, and the actual procedure employed is far from rigorous. Taken literally the computations seem to predict very definite values of bottom velocity. It is surprising that such a 'zero order' theoretical model produces such precise results. A critique of Warren's study is given by Greenspan (1963).

The deep water which flows under the Gulf Stream east of Hatteras most certainly does not flow under the Stream on the Blake Plateau. It also seems likely that this water does not flow northward along the continental rise off the Blake Plateau, but actually joins the Gulf Stream at or slightly north of Hatteras from the east, south of the Stream. Thus we may picture this deep water as flowing slowly toward the continental rise from the east, in a band of latitude from 33° N. to 36° N. Then, upon flowing under the Gulf Stream, it turns abruptly north and then toward the east, and flows as a much more intense current, following the surface Gulf Stream back toward the east again. There is some evidence of the westward flow of deep water south of the Gulf Stream in the slope of deep isotherms upward to the south in fig. 84.

The simple stability theory of meanders presented on page 129 above is incorrect. The lower layer is not treated dynamically; the model is actually a single layer fluid with the value of gravity reduced. Professor N. A. Phillips pointed out to me that a simple translation of axes reduces the whole thing to a state of relative rest.

Stern (1961) has sought a correct formulation of the problem in a reduced gravity model with cross-stream velocity profile. He finds that such streams are stable so long as the cross-stream gradient in potential vorticity does not vanish.

The reduced gravity model has been studied further by Lipps (1963), who, using the data in figs. 32–34, finds a maximum instability for wavelength 180 km., with e-folding amplification rate of 4 days. The meanders observed in Gulf Stream '60 remained at constant large amplitude for more than three months, so obviously do not represent the situation envisaged in the theory. Of course also the passive deep layer implies no interaction with bottom topography. So perhaps the area of pertinence lies east of the Grand Banks of Newfoundland, where the currents actually leave coastal influence.

Duxbury (1963) has produced a meander theory for a two-layer Stream bounded on the left by a resting ocean. He finds stability in the region where the Froude number exceeds unity, the opposite of the incorrect theory given by me on page 129. One would like to see a meander theory built upon a two-layer model with cross-stream velocity profile like that given in equation (12), page 111.

MEANDERS AS A DRIVING AGENCY FOR THE GULF STREAM

Webster (1961) has computed the transfer of momentum against the mean velocity gradient by means of fluctuations observed in the Gulf Stream. Observations are available at two sections across the Gulf Stream south of Cape Hatteras. Similar observations are not available downstream of Cape Hatteras where meanders are more fully developed and of course may therefore act in a different way. In the past, of course, it was supposed that fluctuations represented a transfer of kinetic energy from the energy of the mean stream flow to the eddies and meanders and therefore was essentially of a dissipative nature. Studies of the general circulation of the atmosphere have indicated that the reverse is often true in the air; therefore Webster sought a similar reverse effect in the Gulf Stream where he was able to find a series of measurements of velocity sufficiently long to be significant. The first series of measurements off Onslow Bay has been reported by Von Arx *et al.* (1955) and consists of measurements made by towed electrodes. After making some plausible allowance for the errors introduced by assuming that towed electrodes measure true surface velocity, Webster had available 28 days of measurement as the ship (*Crawford*) sailed back and forth across the Stream. Webster (1961) has given a separate discussion of the hydrography of the meanders measured on this section.

In addition to this set there are also a series in the Florida Straits appearing only in unpublished technical reports of the University of Miami over a four-year period—a total of 632 velocity measurements by towed electrodes from 42 cruises. The number of observations is large by ordinary oceanographic standards, but on the other hand there are not many independent eddies of one-week duration in 28 days, so that the standard error of the means $\bar{u}$, $\bar{v}$, $\overline{u'v'}$ is of comparable magnitude to the means themselves, and they barely emerge from the statistics. Even though this evidence is hardly conclusive, it does point up the important possibility of a previously unexplored mechanism for driving the Gulf Stream, and cautions us about hasty inferences such as appear in page 106 of this work. The largest total transfer of kinetic energy to the mean surface flow occurs off Onslow Bay, where Webster calculates that the latter would be doubled in 11 days. This is a subject which can be investigated in the future when the technique of continuously measuring buoyed current meters has been mastered. At that time it will be possible to gather sufficient data to establish definite statistical significance for such calculations from data.

DEVELOPMENTS OF THEORIES OF WESTERN BOUNDARY CURRENTS

An important point made by Morgan (1956) which was insufficiently described on pages 116 ff. of this work was the fact that a complete description of the wind-driven circulation cannot be made in terms of purely inertial boundary currents alone. The schematic model which Morgan proposed is shown in fig. 93. It consists of three regions: I_i, an

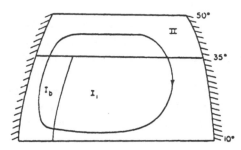

Fig. 93. The three regions of a new ocean model. I_i, interior region; I_b, frictionless stream region; II, northern region. Non-steady and lateral friction effects are possibly important. From Morgan (1956).

interior region where the transport is determined by the Sverdrup interior relation; I_b, a frictionless region on the western side of the ocean where a purely inertial current forms; and II, a northern region of decay of the narrow boundary current where inertia, viscosity, and non-steadiness are all possibly important. Morgan decides to discuss conditions in regions other than II on the assumption that they are independent of a detailed discussion of that region, and that, whatever occurs in II, the flow emerges from there into I_i such that it obeys the interior equations. Morgan's ideas have recently been rediscussed and in one respect corrected, namely, in regard to the existence of eastern boundary currents by Carrier and Robinson (1962) and by Greenspan (1962). Faller (1960) found experimentally that he could produce eastern boundary currents.

In order to limit the discussion here we confine ourselves to a simple exposition of Morgan's ideas due to Veronis (1963a). This paper also contains a discussion of a model ocean, consisting of two superposed layers of different density, which we will not pursue here.

Following Veronis' exposition we take the equations governing the interior of an ocean basin to be

$$-fv = -\frac{1}{\rho}\frac{\partial\rho}{\partial x} + \frac{\partial\tau_x}{\partial z}. \tag{1}$$

$$fu = -\frac{1}{\rho}\frac{\partial p}{\partial y} + \frac{\partial \tau_y}{\partial z}. \tag{2}$$

$$\frac{1}{\rho}\frac{\partial p}{\partial z} = -g. \tag{3}$$

$$\frac{\partial u}{\partial x} + \frac{\partial v}{\partial y} + \frac{\partial w}{\partial z} = 0. \tag{4}$$

$$\frac{d}{dt}(h - z) = 0 \text{ at } z = h, \tag{5}$$

where the quantities are defined as in the earlier portion of the book. Equation (5) defines the free surface height, h.

The equations are integrated over the depth from $z = 0$ to $z = h$. The pressure gradients are expressed in terms of gradients of h. The vertical velocity, w, can be expressed in terms of the horizontal components of velocity at the top and bottom of the ocean through equation (5) and we obtain

$$-fV = -\frac{g}{2}\frac{\partial h^2}{\partial x} + \tau. \tag{6}$$

$$fU = -\frac{g}{2}\frac{\partial h^2}{\partial y}. \tag{7}$$

$$\frac{\partial U}{\partial x} + \frac{\partial V}{\partial y} = 0, \tag{8}$$

where U and V are now horizontal transports defined by $\int_o^h u dz$ and $\int_o^h v dz$, respectively, the stresses at the bottom of the ocean vanish, τ_y evaluated at $z = h$ vanishes, and $\tau_z = \tau$ at $z = h$.

Now cross-differentiating equations (6) and (7) yields

$$V = -\frac{1}{\beta}\frac{\partial \tau}{\partial y}. \tag{9}$$

Substituting V into equation (8) then provides the relation

$$\frac{\partial U}{\partial x} = \frac{1}{\beta}\frac{\partial^2 \tau}{\partial y^2}. \tag{10}$$

Equation (10) does not give unique information about U even when $\partial^2 \tau / \partial y^2$ is known because the zonal transport is determined only to within an arbitrary function of y. It is therefore necessary to introduce higher-order dynamical effects to close the solution.

Assume that quasi-geostrophic balance is valid throughout the interior and up to either the eastern or the western boundary. If $\partial^2\tau/\partial y^2 > 0$ and if equations (9) and (10) obtain up to the *eastern* boundary of a rectangular ocean (where $U = 0$), then $\partial U/\partial x > 0$ and $U < 0$ throughout the interior. A western boundary layer must be added to close the flow. If, on the other hand, $\partial^2\tau/\partial y^2 > 0$ and equations (9) and (10) are assumed valid up to the *western* boundary, then $\partial U/\partial x > 0$ and $U > 0$ throughout the interior. In this case an eastern boundary layer closes the flow.

Veronis shows that there are eight possible flow patterns determined by three independent parameters, namely, the sign of $\partial\tau/\partial y$, the sign of $\partial^2\tau/\partial y^2$, and the validity of equations (9) and (10) up to either the eastern or the western boundary, as shown in fig. 94a to h. In fig. 94a to d quasi-geostrophic flow is assumed right up to the eastern boundary, and a western boundary layer is necessary to close the flow. For example, in fig. 94a the net meridional transport in the interior is northward, and a southward flow must be added in the western boundary layer for purposes of mass conservation. The flow in the boundary layers is assumed to be inertially controlled; i.e., absolute vorticity is conserved along lines of constant transport.

$$\vec{V} \cdot \nabla \left(f + \frac{\partial v}{\partial x} \right) = 0. \tag{11}$$

The local wind stress in the narrow boundary layer region is unimportant.

The meridional velocity essentially vanishes at the place where the interior matches the boundary layer and has a maximum amplitude at the coast. Thus if the flow is northward (southward) in the western boundary layer the relative vorticity, $\partial v/\partial x$, is negative (positive). For the eastern boundary layer northward (southward) flow implies positive (negative) relative vorticity.

Now, applying equation (11) we see that, since $(\partial v/\partial x) + f$ is constant along a streamline, it is necessary that the relative vorticity, $\partial v/\partial x$, compensate for the change in the Coriolis parameter, f, along a streamline. For example, in fig. 1a the flow comes in from the interior and turns southward so that f decreases. For $(\partial v/\partial x) + f$ to remain constant it is necessary that $\partial v/\partial x$ increase, and, since $\partial v/\partial x$ changes from zero to a large positive value, the pattern in fig. 94a is consistent with equation (11).

Similarly, in fig. 94b f increases and $\partial v/\partial x$ decreases along a streamline to yield a consistent picture. On the other hand, in fig. 94c the flow leaves the boundary layer and along a streamline $\partial v/\partial x$ changes from a positive value to zero so that $\partial v/\partial x$ decreases. However, f also decreases and $f + (\partial v/\partial x)$ cannot remain constant but must decrease. Therefore, fig. 94c represents an impossible flow pattern. Similar reasoning shows

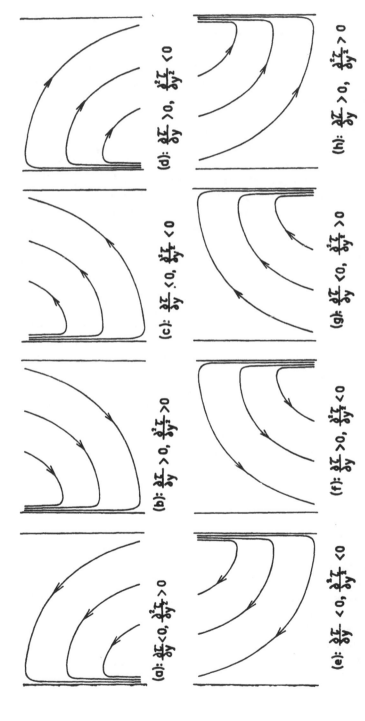

Fig. 94. The eight possible flow patterns in a homogeneous or two-layer ocean. The patterns are determined by the signs of $\partial \tau/\partial y$ and $\partial^2 \tau/\partial y^2$ and by the closure of the flow with either a western (figs. a to d) or an eastern (figs. e to h) boundary current. Some of the patterns cannot satisfy the constraint imposed by conservation of potential vorticity in the boundary layer. From Veronis (1963b).

that the circulations in figs. 94a, 94b, 1e, and 94f are consistent with the necessary constraints and that those in figs. 94c, 94d, 94g, and 94h are not.

In this way Veronis and Morgan are led to the conclusion that *only those patterns with westward flow in the interior satisfy the necessary conditions.*

A particular example is next constructed by Veronis to produce a circulatory pattern first proposed by Carrier and Robinson (1962). Let us consider the wind-stress distribution $\tau \sim - \cos \pi \, y'$ where the non-dimensional coördinate y' ranges from 0 to 1. Here, $\partial \tau / \partial y > 0$ throughout, $\partial^2 \tau / \partial y^2 > 0$ for $0 \leq y' < \frac{1}{2}$ and $\partial^2 \tau / \partial y^2 < 0$ for $\frac{1}{2} < y' \leq 1$. Thus for this example the only possible circulation with inertial balance in the boundary regions is one with fig. 94b for the southern half-basin and fig. 94f for the northern half-basin. This pattern is shown in fig. 95 with a midlatitude jet to provide the eastward flow.

This solution is not unique, the position of the jet at mid-latitude being arbitrary. On the whole, the idea of a mid-latitude jet, flowing across the ocean at the latitude of the maximum wind curl, seems rather unrealistic to the writer. There is a lack of any evidence in the real ocean of such an extraordinary phenomenon. It would be interesting to try to produce something similar in a laboratory experiment, but this may be impossible because of the probable instability of such a long narrow stream. As we will see, in Bryan's numerical model nothing like it occurs. In an unpublished investigation of my own, using several rectilinear vortices to represent crudely the vorticity field, and computing their successive positions numerically, I could see qualitatively that the vortices representing the Gulf Stream on the western side of the ocean easily pass the 'critical' latitude and tend to run around the northern part of the basin before entering a short segment of eastern boundary current. On the basis of this qualitative experience the circulation shown in fig. 95b seems a more likely alternative to that shown in fig. 95a. Fig. 95b agrees better with Bryan's numerically computed patterns, figs. 99 and 100.

The chief physical shortcoming of the pictures drawn above is that there is no provision for the effect of friction which is essential to achieve a steady state. Carrier and Robinson (1962) have attempted to provide frictional effects by introducing viscous sub-boundary layers between the inertial boundary currents and the coasts. This paper is rather intricate structurally. It began as a treatment of a completely inviscid model leading to the circulation illustrated in fig. 95a. During a lecture by G. F. Carrier presenting these inviscid results at the Massachusetts Institute of Technology, N. A. Phillips pointed out the necessity of in-

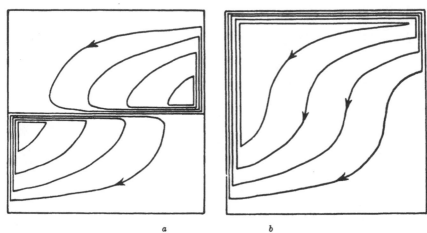

Fig. 95. a. Mid-latitude jet solution. b. Closure by northern rim current.

cluding friction in the model to obtain a steady-state viscous solution. The latter part of Carrier and Robinson's paper is an attempt to meet this objection, but because of the greater difficulty it is not carried out in the same deductive manner as the first part, and therefore remains an attempt to justify retention of all possible features of the inviscid model with viscosity and turbulence treated in a heuristic manner. No other paper has attempted to treat such high Reynolds number flows (except, of course, Fofonoff's free solutions). The hypothetical picture of the ocean circulation which features the mid-latitude jet (fig. 95a) differs from the Munk picture (fig. 60) not only in details of arrangement of the boundary currents but also because, in place of a single large sub-tropical gyre (between latitudes 14° N. and 50° N.) centered about the latitude of the maximum curl of the wind-stress (at 33° N.), it features two gyres, with a boundary at that latitude, and each gyre of different east-west asymmetry. If we also treat the subpolar gyre in the same way we could generate perhaps a total of four gyres which seems to be too many to fit the actual distribution of dynamic topography of each major ocean basin.

Figs. 96a, b, and c show the mean wind-stress, the curl of the mean wind-stress, and the dynamic topography of the sea-surface, of the world ocean, taken from a summary paper by Stommel (1964) in the Hidaka Jubilee Volume. The reader is referred to the original paper for details of construction of these charts. The main point here is to offer them for comparison with the various theoretical flow patterns described.

The problem of producing a thoroughly satisfactory theoretical model of the wind-driven circulation has not been solved. One way of producing

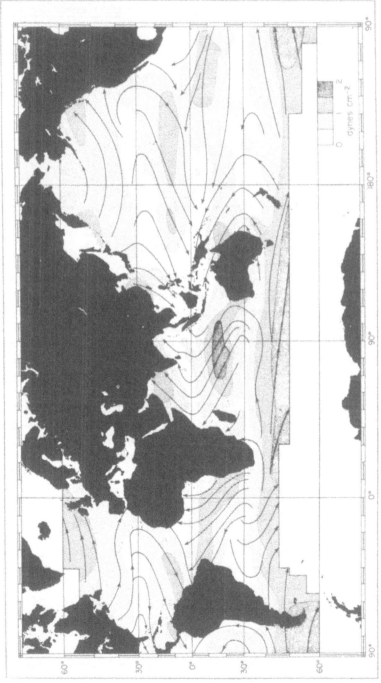

Fig. 96a. The mean annual stress of the wind at the sea surface, drawn from the tables of Hidaka (1958).

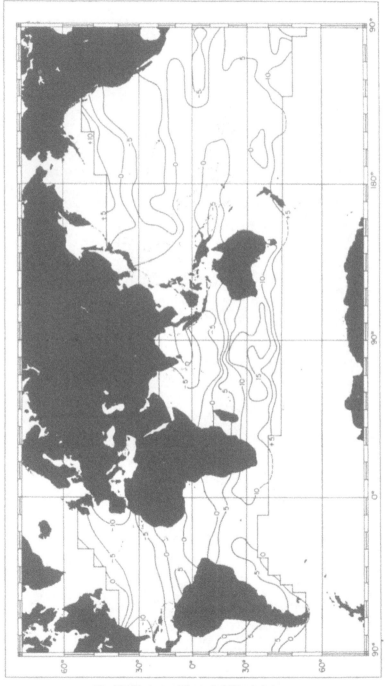

Fig. 96b. Contours of the curl of the wind-stress.

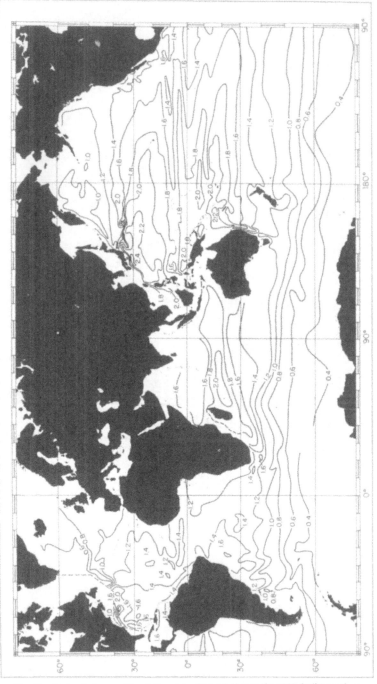

Fig. 96c. Dynamic topography of the sea surface relative to the 1000-decibar surface.

better agreement with observation would be to consider a model in which the Reynolds number is not uniform: low viscosity in the equatorward half-basins of subtropical gyres[1], and large eddy viscosity in poleward halves. This is a rather distasteful last resort from a theoretical point of view, and has not been used explicitly in any models; it has only been hinted at.

If we consider Munk's solutions as very small Reynolds number solutions, and the mid-latitude jet model of Carrier and Robinson as an extremely large Reynolds number possibility, then we see that there is a wide range of intermediate Reynolds number which needs exploration. Munk, Groves, and Carrier (1950) have explored small Reynolds number (less than unity) circulation patterns. Moore (1963) and Ilyin and Kamenkovich (1963) have attempted to construct models for Reynolds number near unity.

Moore (1963) proposes to do away with boundary layers altogether in the northern half-basin where all the difficulty is encountered, and to allow a wave-like (instead of exponential) solution starting with large amplitude at the western boundary and extending eastward, slowly being damped by viscosity—a train of damped Rossby waves extending into the interior. An example worked out by Moore is shown in fig. 97. There

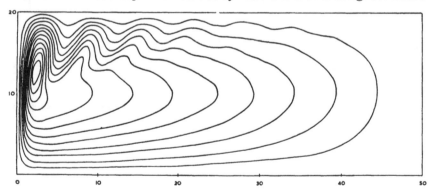

Fig. 97. Lines of constant $(\beta L'/W_o \pi) \psi$ for rectangular basin with $L = 5000$ km. and $L' = 2000$ km. The Reynolds number $U^{3/2}/\beta^{1/2} = 5$. From Moore (1963, fig. 1).

are some approximations in the derivation of equation (a form of linearization of the wave-equation) used in deriving the waveform which have not been completely justified. The form of Moore's solution changes with a Reynolds number defined as $U^{3/2}/A\ \beta^{1/2}$ where A and β have the familiar definitions of Munk's theory (Chapter VII) and U is

[1] Despite its widespread use, there does not appear to be such a word as 'gyral'. But there are gyres and spirals.

the amplitude of the smoothed mean east-west component of velocity in the interior of the basin. The viscous theories of Chapter VII correspond to the limit as the Reynolds number vanishes; the pure inertial theories of Chapter VIII correspond to infinite Reynolds number (at least, those parts outside the decay region do), as do also the circulations discussed by Greenspan, Carrier and Robinson, and Veronis. Fig. 97 is computed for Reynolds number 5. Scaled to the North Atlantic this corresponds to an eddy viscosity of 5×10^5 cm.²/sec. If the Reynolds number is made larger, the amplitude of the waves in the wavy portion of the figure increases. If the Reynolds number decreases, we move over smoothly to Munk's gyre with north-south symmetry.

Bryan's numerical model of a western boundary current.—Bryan (1963) has carried out numerical time-integrations of the wind-driven ocean model using a homogeneous ocean, a rigid lid on the top to filter out gravitational-inertial waves, and the assumption that the driving force of the wind and frictional dissipative terms are all independent of depth. He obtains solutions which help to shed much light on the actual form of solution that the homogeneous ocean will assume, and how the Gulf Stream should theoretically conduct itself after it passes the latitude of maximum wind curl. The equation which Bryan uses is

$$\frac{\partial}{\partial t} \nabla^2 \psi + R_o \left[\frac{\partial}{\partial y} \left(\frac{\partial \psi}{\partial x} \nabla^2 \psi \right) - \frac{\partial}{\partial x} \left(\frac{\partial \psi}{\partial y} \nabla^2 \psi \right) \right] + \frac{\partial \psi}{\partial x}$$
$$= - \sin \frac{\pi}{2} y + \frac{R_o}{R_e} \nabla^4 \psi, \qquad (12)$$

where the various quantities have been reduced to dimensionless form in the following way:

$$x' = Lx,$$
$$y' = Ly,$$
$$\psi' = \rho_o h L V_b \psi,$$
$$t' = (\beta L)^{-1} t,$$
$$V_s = \frac{w_o \pi}{2 \beta L \rho_o h}.$$

Here we have reversed Bryan's notation, using primes to indicate the original variables, and the unprimed to represent the new dimensionless ones. L is the horizontal scale of the oceanic basin, w_o the amplitude of the wind stress, and h the depth of the ocean if we assume it to be homogeneous (Bryan does not, and is led into complicated ratiocinations amounting to much the same ultimately). The other quantities have been defined before. V_s is a scale velocity associated with the interior Sverdrup flow. R_o is a Rossby number defined by $R_o = V_s/\beta L^2$, and the Reynolds number in this case is defined by $R_e = (V_sL)/A$. The scale length, L, is comparable to the earth's radius; therefore βL is equivalent to the Coriolis parameter. In this case R_o may be interpreted as a Rossby number for the north-south flow in the interior of the basin. Since the depth is uniform, continuity assures that the scale velocity of the southward flow in the interior multiplied by the width of the basin is approximately equal to the scale velocity of the return boundary current, $V_{\text{B.C.}}$, multiplied by the boundary current width, d,

$$R_e = (V_sL)/A \simeq (V_{\text{B.C.}}d)/A.$$

Thus R_e can be interpreted in the conventional way as a Reynolds number of the boundary current.

TABLE 14

VALUES OF ROSSBY NUMBER R_o FOR β EQUAL TO 10^{-13} SEC.$^{-1}$ cm.$^{-1}$ AND w_o EQUAL TO 2 DYNES cm^{-2}
$2L$ is length of rectangular basin in north-south direction

Depth h	2L	
	5000 km.	10,000 km.
200 m.	1.00×10^{-3}	1.24×10^{-4}
400	0.50×10^{-3}	0.62×10^{-4}
800	0.25×10^{-3}	0.31×10^{-4}

TABLE 15

VALUES OF THE REYNOLDS NUMBER R_e FOR VARIOUS VALUES OF A AND h
β and w_o have same values as in table 14

Depth h	A	
	5×10^7cm.^{2}sec.$^{-1}$ (Munk, 1950)	10^8cm.^{2}sec.$^{-1}$ (Stommel, p. 107 above)
200 m.	31	1569
400	16	863
800	8	431

Tables 14 and 15 give values of R_o and R_e for different choices of the depth of a surface layer h, and the length of the rectangular basin in the north-south direction. In order to give a reasonable value for the transport in the wind-driven gyre in spite of the narrowness of the basin, the amplitude of the wind-stress function is taken to be 2 dynes cm.$^{-2}$. This is approximately two times the representative value for the Northern Hemisphere oceans (Hidaka, 1958).

The transport corresponding to the values of β and w_o used in table 14 is 31.4×10^6 metric tons sec^{-1}, or about 1/2 to 2/3 the estimated transport of the Gulf Stream. Bryan integrates equation 12 from rest numerically. We will not go into the important details of the numerical integration here, nor can we answer such critical questions as whether enough grid points fall in the boundary current to give an adequate representation of the non-linear processes there, but will merely refer the reader to the original paper. Initially there is á transient rise of the kinetic energy of the system, and then oscillation about a 'steady state' value. For large Reynolds numbers the oscillation does not die out with time, and this is one of the limitations of the computational method. However, for lower Reynolds numbers the oscillations are damped after several cycles and only small fluctuations remain. Bryan shows that these fluctuations may be identified as Rossby waves in the basin.

We are primarily interested, however, in comparing these numerical solutions with those obtained by various analytical methods in earlier work. For this we must consider the near-equilibrium solutions obtained after the computation has steadied.

To gain insight into the effect of the two basic parameters, R_o and R_e, different cases were worked out by Bryan for the rectangular basin and a wind stress proportional to $\sin(\pi y/2)$. The values of the Rossby number and the Reynolds number for each case are shown in fig. 98. The

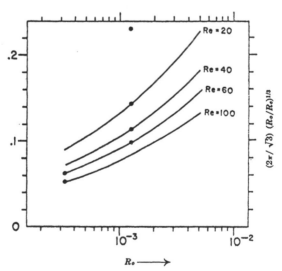

Fig. 98. The width of the boundary current based on Munk's steady-state linear theory is the ordinate, and the Rossby number is the abscissa. The dots indicate values of the parameters for which numerical solutions were obtained. From Bryan (1963, fig. 3).

ordinate of fig. 98 is the ratio of the width of a purely viscous boundary current to the total width of the basin. The solutions indicate that the actual inertial-viscous boundary layer of the non-linear solutions is wider than the purely viscous boundary layer given by Munk's solution in the southern half of the basin, and narrower in the northern half, but the average width is very similar.

In order to show the gradual transition from a purely viscous regime to a regime in which non-linear effects are important, Bryan exhibits a series of cases (fig. 99) for R_o equal to 1.28×10^{-3} and R_e equal to 5, 20

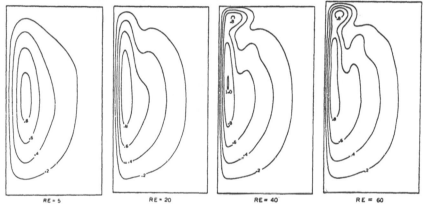

Fig. 99. Patterns of the transport stream function, ψ, obtained from a time average of solution with $R_o = 1.28 \times 10^{-3}$ and four values of R_e. For $R_e = 5$ and 20, the finite difference interval was $\Delta S = 1/20$; for $R_e = 40$ and 60, $\Delta S = 1/40$. From Bryan (1963, fig. 4).

40, and 60. One sees that, as the Reynolds number is increased (here defined differently from Moore), the circulation goes over from a simple north-south symmetry to a circulation with narrower boundary current, and wavelike formations at the top. These are similar to Moore's picture, or alternatively may be described as a countercurrent which feeds water back westward toward the Stream and therefore permits it to penetrate further north than the latitude of the maximum of the curl of the wind-stress. No mid-latitude jet forms at all.

Fig. 100 shows the flow patterns for two cases in which the Reynolds number is kept constant, and the Rossby number varied by a factor of 4 from 1.28×10^{-3} to 3.2×10^{-4}. A constant Reynolds number means that inertial and viscous effects retain the same ratio. For the particular scaling used in this study, the beta effect in the interior is always of order unity. Thus when the Rossby number is decreased the relative importance of the beta term is increased. As a result, the viscous-inertial boundary current becomes increasingly concentrated with low Rossby

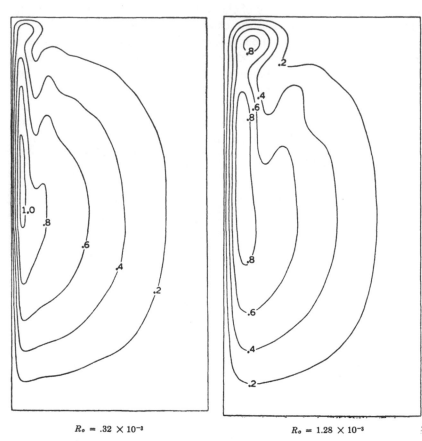

$R_o = .32 \times 10^{-3}$ $R_o = 1.28 \times 10^{-3}$

Fig. 100. Two patterns of the transport stream function for a fixed Reynolds number of 60 and two values of the Rossby number are given to illustrate the effects of varying the latter. From Bryan (1963, fig. 6).

number. Fig. 101 shows the magnitude of the various terms of the vorticity equation for the case $R_e = 40$, $R_o = 1.28 \times 10^{-3}$. Although the current illustrated was calculated for $R_e = 40$, the reader will notice that the viscous terms are at least three times the inertial in this boundary layer, and this seems to suggest that physically the Reynolds number is rather nearer 0.3 than the larger value from formal definition. One of the most interesting features is that the northern half of the boundary current is almost entirely viscous, while the southward-moving counter-current has an inertial-viscous character similar to the southern portion of the boundary current.

The case for $R_o = 3.2 \times 10^{-4}$ and $R_e = 100$ was examined in greater detail because it represents the extreme of low Rossby number and high Reynolds number for which it was feasible to make a computation. The maximum Reynolds number is limited by the very dense numerical grids that must be used to define adequately the boundary current. Fig. 102a shows the variation with respect to time of the kinetic energy. The difference interval used for this computation is 1/60. Another case for a Reynolds number of only 50 is shown for comparison. In both cases initial fluctuations are set up by transient Rossby waves. If the calcula-

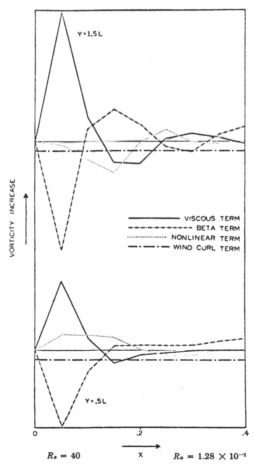

Fig. 101. The variation of different terms of the vorticity equation with distance from the western boundary for a solution approaching equilibrium with $R_o = 1.28 \times 10^{-3}$ and $R_e = 40$. From Bryan (1963, fig. 7).

tion is restarted with the average values over the interval I (fig. 102a), it remains steadier, as shown by the interval II. Further details of the instability of the boundary current are shown in fig. 102b, c, and d.

Bryan has tested the effect of having an irregular wall along the western boundary and finds that the current does not separate easily from sharp bends in the wall, but sometimes even builds an intensified transport at the bend by making an extra eddy just outside the current, thus increasing the transport. Bryan also computes some examples with more realistic distributions of wind-stress than the simple schematic zonal winds most models have employed. He finds that there is a tendency for the boundary current to separate from the coast at the latitude of minimum wind-stress curl instead of the maximum, and this is perhaps the most realistic picture which theory has produced to date.

In the laboratory it is possible to produce a regime analogous to that envisaged in the above theoretical investigations by driving the fluid in a homogeneous rotating fluid of non-uniform depth by distributed sources and sinks (Stommel et $al.$, 1958; Faller, 1960). Many interesting flow patterns, which are in accord with the ideas presented here, have been produced, and the effects of various placements of barriers have been qualitatively investigated. Some of the flow patterns which occur are so counter to the dictates of everyday non-rotating intuition that it is very reassuring to see that they actually do occur in these model experiments.

Non-steady solutions with a steady applied wind-stress. — Veronis (1963b) has embarked upon an ambitious series of studies of numerical models of the oceanic circulation using truncated Fourier series representation of the transport field instead of trying to represent it by a finite difference grid. The study is not yet completed, but he has published a first study for the extremely crude representation which one obtains with only two Fourier components in each direction. Even this primitive study exhibits the most surprising and unexpected effects. Veronis's basic equation,

$$\frac{\partial}{\partial t} \nabla^2 \psi + R_o \left[\frac{\partial \psi}{\partial x} \nabla^2 \frac{\partial \psi}{\partial y} - \frac{\partial \psi}{\partial y} \nabla^2 \frac{\partial \psi}{\partial x} \right] + \frac{\partial \psi}{\partial x} = - \epsilon \nabla^2 \psi - \sin x \sin y, \quad (13)$$

differs from Bryan's only in that he uses a frictional term similar to that used on page 88 above. The quantity ϵ is simply $R_o/R_e \equiv R/\beta L$ where R is the frictional coefficient of page 88.

If we introduce the following extremely crude representation of the stream function,

$$\psi(x, y, t) = a(t) \sin x \sin y + b(t) \sin 2x \sin y + c(t) \sin x \sin 2y$$

$$+ d(t) \sin 2x \sin 2y,$$

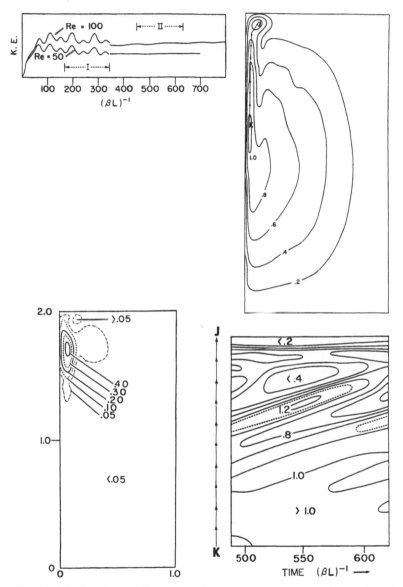

Fig. 102a. The time change of kinetic energy for $R_o = 3.2 \times 10^{-4}$ and Reynolds numbers of 50 and 100. Fig. 102 from Bryan (1963, fig. 8).

Fig. 102b. The pattern obtained by averaging the $R_e = 100$ case over the interval indicated by II in fig. 102a.

Fig. 102c. The pattern of the r.m.s. deviation from the time average over the period indicated by II in fig. 102a ($R_e = 100$).

Fig. 102d. The change of the amplitude along the line K–J of fig. 102b with respect to time ($R_e = 100$).

and substitute this into the hydrodynamical equation, keeping only terms listed above, and dropping all others, we obtain the following equations for the time-derivatives of the four Fourier coefficients:

$$\dot{a} = -\frac{4}{3\pi} b - \epsilon a + \frac{1}{2},$$

$$\dot{b} = \frac{8}{15\pi} a + \left(\frac{9}{20} R_o\right) ac - \epsilon b,$$

$$\dot{c} = -\frac{8}{15\pi} d - \left(\frac{9}{20} R_o\right) ab - \epsilon c,$$

$$\dot{d} = \frac{1}{3\pi} c - \epsilon d.$$

(14)

The first terms on the right-hand side are the β terms; the terms with R_o factor are the non-linear terms; with factor ϵ are frictional terms; and the remaining $1/2$ is the driving term. If we replace the $1/2$ by zero, and set $\epsilon = R_o = 0$, then we have two simple harmonic oscillations: linear Rossby waves in the basin. If we now restore R_o to a non-zero value, but set the left-hand derivatives equal to zero, we find that we cannot obtain Fofonoff's full non-linear solution by so limited number of terms.

Now let us return to the full four equations and seek steady solutions. Formally, we set the time derivatives equal to zero, eliminate b, c, d, and obtain a cubic for a. For small ϵ and R_o, in which case we find only one real solution,

$$a \approx 0, b \approx 3\pi/8, c \approx 0, d \approx 0,$$

which corresponds—as well as such crude representation can—to the linear solution shown in fig. 58. For large ϵ we obtain essentially

$$a = \frac{\epsilon^{-1}}{2}, b \approx \epsilon^{-2}, c \approx \epsilon^{-3}, d \approx \epsilon^{-4},$$

which is similar to fig. 55, and this is true regardless of the value of R_o.

However, if we now hold $\epsilon << 1$ but raise R_o past 0.32, we discover that a transition from one to three real roots occurs, and discussion of the stability of the solution becomes complicated. Veronis used an analog computer to see which solutions the system will go to if started from rest in this case and found that the solution with a dominant is stable instead of that with b dominant. But an even more remarkable behavior is found: for the case $0.22 < R_o < 0.32$, where only one steady solution exists (*b*-dominant), a system started from rest does not settle down to the steady solution, but goes into a non-steady limit cycle which in phase space describes an orbit encircling the *b*-dominant solution and the position where the *a*-dominant solution would be found were $R_o = 0.32$. The time spent during the cycle is greatest in the neighborhood of the *b*- or *a*-dominant solutions, depending upon whether R_o is near the lower or upper limit of the range $0.22 < R_o < 0.32$, respectively. Veronis is now proceeding to a study of representations with many more Fourier terms.

A useful collection of reprinted papers on the wind-driven circulation has been produced by Blaisdell Press (editor, A. R. Robinson, 1963). The selection follows essentially the sequence of papers referred to in chapters 7 and 8 of this book, plus the Carrier and Robinson (1962) paper. The interesting papers by Greenspan, Bryan, Moore, and Veronis are not included. Fofonoff (1962) has produced a very useful and interesting review article on the dynamics of ocean circulation, which is highly recommended to anyone trying to find his way into the theoretical literature.

An idea implicit in much of this book, explicit on page 154, and made use of by Stommel and Arons (1960) is that, in order to construct circulation patterns in the ocean, one can always locate the higher-order dynamics on the western side of the ocean, and compute the form of the interior on the basis of geostrophy and mass conservation alone. This is of course a rather unsophisticated idea and not proven by a general rigorous deductive line of reasoning. In the case where viscosity dominates it does seem to be reasonably proved, but when the Reynolds number is increased and inertial effects become dominant no one really knows what the current pattern should be. The present state of knowledge of high Reynolds number flow as presented by Carrier and Robinson (1962) is suggestive of difficulties, but inadequate to determine unique solutions. Do inertial boundary layers become unstable? Do they interact frictionally with the bottom through meander formation? Indeed, is the very idea of a boundary current relevant in the decay regions of the oceanic circulation?

In the theories of Moore (1963) and Ilyin and Kamenkovich (1963) and in the numerical studies of Veronis (1963*b*) and Bryan (1963), we are witnessing a breaking away from an overenthusiastic application of singular perturbation theory, and in the planning for further observational study of the interior there is a growing awareness of the importance of transient processes, particularly in the decay regions.

In nature we do not seem to encounter mid-latitude jets, nor inertial eastern boundary currents that are important parts of the large-scale grand circulation pattern. The higher-order effects do seem to be con-centrated on the western sides of the great oceanic gyres, whether we can account for this concentration completely satisfactorily or not.

I think that we would be more profitably employed if we puzzled less about why the boundary layer separates from the coast somewhere and more upon why the ocean current's behavior departs from that of a boundary layer.

THE THERMOHALINE CIRCULATION

There has been much progress in the theory of the oceanic thermocline and the associated thermohaline circulation. The subject of Chapter XI therefore has grown into a topic in its own right, and since it deals with the open ocean far from coasts and boundary currents it really is not the province of this book. The two early papers on this subject, now rather thoroughly superseded, are the paper by Lineykin (1955), which was the first attempt to deal with a continuous density distribution by intro-ducing a simplified density transfer equation, and the one by Stommel and Veronis (1957), where the effect of the variation of Coriolis parameter with latitude was shown to determine the scale-depth of the thermocline. These are both discussed in a review article by Stommel (1957; see note 4, page 172, above). These papers suggested the mathematical problem, and the search for technique of solution was carried further in two papers by Robinson and Stommel (1959) and Welander (1959), the first of which dealt with a linearized model with mixing, the second with a non-linear model without mixing. The final model, non-linear with mixing, was successfully formulated, using similarity transformations, by Robinson and Welander (1963) in a magnificent piece of work. Numerical solutions of this equation for a particular case have been provided by Stommel and Webster (1962) and analytical solutions for another special case by Blandford (in press). The basic dynamical equation outside the Ekman boundary layer is equation (27) on page 170. Coupled with this is an equation expressing the transport of density by advection and diffusion. The general patterns of circulation are the same as described qualitatively in Chapter XI, but of course the advantage of the mathematical solu-tions is that they give quantitative relations between the depth of the thermocline, the intensity of vertical mass flux beneath the thermocline, the thickness of the thermocline, the eddy conductivity, and so on. The solutions apply only in the open ocean away from coasts. The theory of the higher-order boundary currents that are needed to close off such models at the coasts (analogous to the boundary currents involved in theories of homogeneous oceans) has not yet been developed.

Application of some of these ideas to patterns of abyssal flow, refining ideas underlying fig. 82, has been made in a series of three papers (Stommel *et al.*, 1960).

RECOMMENDATIONS

Looking back upon the recommendations at the end of Chapter XII, I realize now that they were rather inadequate when put up against the complexity and difficulty that really underlies the problem of deepening our knowledge of the Gulf Stream by planned observations. Certainly the experience of the past years has shown us that the structure of deep currents is much more complex than envisaged in 1957, and that statistically significant information is going to be difficult to come by. The actual pattern of deep flow along the continental shelf and under the Gulf Stream is not clearly defined even six years after the pioneering measurements of Swallow and Worthington. Perhaps we should be grateful for the limited visibility into the future which shields us from recognizing the full burden of future work which we still have before us. I have written a short article (Stommel, 1963) describing the difficulties which oceanographers have encountered in the past in trying to design fruitful programs of observation at sea.

There is a danger, in trying to see what ought to be done, of adopting an excessively narrow point of view: in a subject such as the Gulf Stream there is a need for the broadest and multicomponent mode of approach. Thus we need more descriptive-geographical surveys, such as Gulf Stream '60. We need detailed studies of such features as the cold eddies so that we can describe them in the same kind of "case history" detail as such meteorological scientists as Palmén do so beautifully for analogous atmospheric phenomena. We need long-distance tracking of deep floats by sofar; we need diffusion experiments on all scales; we need measurements of current by bottom-mounted instruments. We need to study the Gulf Stream from many points of view: from that of the geographer and that of the applied mathematician. We must not be carried away with some momentarily popular but restricted idea such as power spectrum analysis or the language of homogeneous turbulence or theories of homogeneous steady-state models to the extent that we neglect other interpretations. With these reservations and animadversions, I think that we can nevertheless see ways in which the future observational programs may usefully develop. One of these ways is likely to be opened by the development of buoy technology which eventually will provide us with means to measure velocity and other properties continuously at many points in the ocean for long periods. Considering the technical difficulties being encountered, it may be 1970 or 1975 before really reliable techniques and instruments are perfected. However, we can consider at present the kind of results we will look for. In short, we will want to investigate the role of transient eddy processes in the ocean; in other words, we must come to terms with the ocean as a large-scale turbulent medium and design our programs of measurement to reveal the nature of the turbulent processes.

In order to elucidate why it is crucial to obtain quantitative information on the transport of momentum and vorticity by large-scale eddy processes in the ocean, we call attention to an analogous position in the development of the theory of the general circulation of the atmosphere about 1953, when the fundamental quantitative studies of the statistics of upper air data carried out by Starr and his collaborators overthrew the classical picture of a predominantly meridional circulation in which it was thought that the observed large-scale fluctuations played a more or less passive dissipative role. Starr's investigations of the observations demonstrated that the fluctuations actually drive the mean circulation, and present-day theoretical studies of the atmospheric circulation all allow the fluctuations to play this more important role.

We are in a similar position in oceanography. The fundamental concept, about which all the theoretical investigations from 1947 to 1962 are pivoted, is the basic Sverdrup relation between the local curl of the mean wind-stress and the vertical integral of the meridional component of velocity. The theory of the thermocline, of the thermohaline, and of wind-driven circulation all depend upon this simple idea: that large-scale, quasi-geostrophic eddy processes do not play an important dynamical role in the vorticity balance in the interior of the ocean.

During the past few years serious doubts about the neglect of eddy processes have begun to arise:

(a) The *Aries* measurements in the Atlantic, originally planned by Swallow, Crease, and Stommel to determine the mean velocity field at different depths, unexpectedly revealed the presence of large-scale long-period eddies whose r.m.s. amplitudes were two orders of magnitude greater than the expected means, indeed so large that it is difficult to imagine that they can be decoupled from the mean fields as is implicit in the Sverdrup relation. At any rate the irregular motions were so large that it was not possible to test the Sverdrup relationship in the simple way which the *Aries* measurements were originally intended to do. In order to obtain a statistical description of these eddy processes and to be able to map and describe them, it is evident that an effort at least an order of magnitude greater than the *Aries* measurements is necessary.

(b) Calculation of the amplitude of the abyssal circulation from I.G.Y.

and Norpac data by the method of Stommel (1956; see note 4, p. 172 above) yields abyssal circulation rates much too large to be compatible with water mass analysis and radiocarbon data. The same lack of agreement appears when the thermocline theory is semi-quantitatively applied to the actual distribution of density in the ocean. These discrepancies also suggest that something important is left out of the simple Sverdrup relation.

(c) Various simple theories of baroclinic instability (e.g., Phillips, 1951), when applied to the laminar thermocline theories based on a laminar interior regime (following Sverdrup), indicate that the interior solution as given by the thermocline theories of Stommel, Robinson, and Welander is dynamically unstable. But because of the immense complication of the theoretical problem of computing the fully developed geostrophic turbulent processes in the oceanic thermocline and the very incomplete observational description of such processes, it is not possible to develop the theory of the unstable thermocline further at present. When more observational guide lines are available it seems probable that the theory can proceed, numerically if necessary. Of course it is not at all clear whether the important property-transferring eddies owe their existence to instabilities of the thermocline or of the coastal boundary currents, or are induced by irregularities of bottom topography or of the applied wind-stress.

(d) Early theories of the oceanic circulation (Rossby, Hidaka, Stockmann) placed much emphasis on the hypothetical existence of large lateral eddy-transports. Sverdrup banished them from the open ocean, and Munk found that their influence might be limited to the western sides of oceans, and was able to compute fields of transport in the ocean which bear considerable resemblance qualitatively to the observed geographic mean distribution of ocean currents. The magnitude and role of the eddy processes envisaged in the Munk theory are purely hypothetical. At present only extremely rough evaluations of eddy processes in the Gulf Stream itself have been possible, that is, Stommel (1955b) on the basis of Pillsbury's data, and Webster (1961a) on the basis of Von Arx's data. In the open ocean there may be a turbulent flux of vorticity by eddies which quantitatively modifies the vorticity balance in the interior expressed by the Sverdrup equation.

Estimates of the importance of non-linearities in the open ocean have until recently been made in terms of the amplitude and scale of the mean motion, but the Swallow-Crease results indicate the turbulent terms to be the important ones. If we take a mean flow of 1 cm./sec. the β term is $O(10^{-13})$ (per unit depth), while if the r.m.s. fluctuation is 10 cm./sec. with a horizontal scale of 100 km., the turbulent term is $O(10^{-12})$ times the correlation coefficient. Perfectly correlated eddies of this type would completely destroy the Sverdrup balance, while a correlation of 1/10 would imply approximately equal contributions. It is important to note that mean or Reynolds non-linearities may affect the circulation through the vorticity balance even though the momentum balance is essentially geostrophic.

Before the end of the 1960's it may be possible to launch a large-scale buoy program, designed to measure the pertinent statistical properties of geostrophic eddies, this program to cover a period of several years.

With analysis of the Richardson buoy data we will be better able to define the first step in this experiment. To date, the only information about irregular motions in the ocean off Bermuda is that of Swallow and Crease. In summary, they find periods of several weeks, a minimum scale of about a hundred miles, and amplitudes of about 10 cm./sec. Although the Richardson buoy data may somewhat modify these numbers, it is clear that in order to obtain full spatial and temporal spectra, several scales of separation of buoys are needed in the network. The problem of conducting an observational study of this magnitude is formidable. When we recall further that the buoy program is only one part of a many-sided approach to studying the ocean circulation, we have a measure of the magnitude of the task before us.

Appendix One

GLOSSARY OF SYMBOLS

The definitions of only the chief mathematical and physical symbols used in this book are indicated below. For a subscript or mark appended to a symbol and used only once or twice, no special listing is given.

SYMBOLS FROM THE ROMAN ALPHABET

A average coefficient of horizontal-eddy viscosity

a, b, c, d in Chapter XIII, Fourier coefficients

b north-south dimension of a rectangular basin

c internal long-wave velocity, $\sqrt{(g'h_0)}$

D undisturbed depth of a homogeneous ocean, or of the top layer of a two-layer ocean

F body force representing wind stress at surface; also, in Chapter VIII, a function of ψ

f Coriolis parameter

G a function of ψ in Chapter VIII; total transport of western current in Chapter XI

g acceleration due to gravity

g' reduced gravity, equal to $g(\Delta\rho/\rho)$

H intensity of the earth's magnetic field; subscripts denote components

h perturbed depth

K_H horizontal coefficient of eddy viscosity

K_V vertical coefficient of eddy viscosity

k wave number

L horizontal length scale

l length of cable between towed electrodes

M_x, M_y horizontal components of vertically integrated mass transport per unit width

p pressure

R an artifice, coefficient of friction on the bottom of a homogeneous ocean, assuming a linear dependence on velocity

R_e Reynolds number

R_o Rossby number

$\mathscr{R}$ radius of curvature of a streamline

r east-west dimension of a rectangular basin

s north-south dimension of wind system

T transport per unit depth

$\mathscr{T}$ vertically integrated transport

T-S   the T-S diagram, with temperature as ordinate, salinity as abscissa; used in water-mass analysis

U mean velocity, used to linearize inertial terms in the perturbation theory

U, V in chapter XIII horizontal components of vertically integrated velocity

u horizontal component of velocity in the x-direction

v horizontal component of velocity in the y-direction

x horizontal coördinate, usually meaning (in this book) directed eastward

y horizontal coördinate, usually meaning (in this book) directed northward

z vertical coördinate, directed upward

SYMBOLS FROM THE GREEK ALPHABET

$\alpha \;=\; 1/\lambda$

β meridional gradient of the Coriolis parameter

γ^2 drag coefficient of wind on sea surface

$\epsilon \;=\; R/\beta L$

ζ relative vorticity, vertical component

λ radius of deformation

μ	wavelength unit (microns) of radiation
ν	meander frequency
ρ	density of water
ρ_a	density of air
σ =	$10^3(\rho-1)$
σ_t	a special measure of density, defined on p. 31
τ	shearing stress across a horizontal surface
ψ	stream function, or mass-transport function, depending on the context

SPECIAL SYMBOLS

‰	parts per thousand, unit of salinity
*	star, used to denote quantities in inertial boundary layer
→	arrow, used over a quantity to indicate a vector

Appendix Two

SOURCES OF DATA

HYDROGRAPHIC STATION DATA

All hydrographic station data obtained in the Gulf Stream, consisting of precise temperature and salinity measurements, have been published at Copenhagen since 1910 by the Conseil permanent international pour l'exploration de la mer in its annual *Bulletin hydrographique*, for the period beginning with the year 1908. Because of the great care taken in editing the *Bulletin*, there is an interval of about five years between the date of observation and the date of publication.

A complete index, including charts, of all the hydrographic data from the Atlantic Ocean prior to 1932 is contained in *International Aspects of Oceanography*, by T. Wayland Vaughan (1937).

Two papers by C. O'D. Iselin (1936, 1940) may be used as an index for *Atlantis* stations made in the Gulf Stream between 1930 and 1940; the original data are to be found in *Bulletin hydrographique*.

Since the Second World War there have been relatively few hydrographic sections of the Gulf Stream; specifically:

1946, Sept. *Atlantis* stations 4414–4423 (no salinities), a Chesapeake Bay–Bermuda section.

1947, June. *Atlantis* stations 4565–4570, a section off Cape Hatteras.

1947, Dec. *Atlantis* stations 4623–4627, a section along 65° W. longitude.

1949, July–Sept. *Atlantis* stations 4760–4832, an areal survey near the Grand Banks.

1950, June. International Ice Patrol stations 4175–4184, a section near 50° W. longitude (see Soule, 1950).

1950, Oct.–Nov. *Atlantis* stations 4842–4882, three sections at 68° W. longitude (see Worthington, 1954 b).

1953, Aug. *Atlantis* stations 5081–5125, three sections at about 73, 70, and 61° W. longitude respectively.

1954, Sept. *Atlantis* stations 5176–5202, along 65° W. longitude, a very deep section.

1955, June. *Atlantis* stations 5295–5312, Chesapeake Bay to Bermuda, very deep.

1955, June–July. *Atlantis* stations 5344–5360, east-southeast of Cape Romain, S.C., very deep.

1956, Nov. *Atlantis* stations 5416–5444, south of Grand Banks on 50° W. longitude, very deep.

The data from all these hydrographic stations have been submitted to the *Bulletin hydrographique*, and have been or will eventually be published.

BATHYTHERMOGRAPH DATA

The data from bathythermograph measurements are so voluminous that they have never been published in full. The information is available only in raw, unreduced form in the card files of the Woods Hole Oceanographic Institution, and in the form of rough running sections drawn aboard ship. A few sections have been published (see Iselin and Fuglister, 1948). S. L. Strack has made an analysis of fifty-eight bathythermograph crossings of the Stream to study the relation of temperature at the surface to that at various depths as a function of season, but this study has not been published.

DISCUSSIONS AND SUMMARIES OF ATLANTIC OCEAN DATA

In addition to the sources mentioned above there are several general studies, atlases, chart collections, and so on, containing much useful information about the Atlantic Ocean, but not concerned specifically with the Gulf Stream.

1. Charts of surface temperature, salinity, and density, according to season, are given in Günther Böhnecke's (1936) atlas, consisting of seventy-four plates and entitled *Atlas zu: Temperatur, Salzgehalt und Dichte an der Oberfläche des Atlantischen Ozeans*, in *Deutsche Atlantische Expedition auf dem Forschungs- und Vermessungsschiff* Meteor, *1925–1927, Wissenschaftliche Ergebnisse*. (Band 5—Atlas.)

2. Charts of distribution of mean temperature, salinity, and density at various depths are included in an atlas of 103 plates prepared by Georg Wüst and Albert Defant (1936), *Atlas zur Schichtung und Zirkulation des Atlantischen Ozeans*, which is also in the multivolume work on the scientific results of the *Meteor* Expedition (Band 6—Atlas).

3. For charts of the distribution of properties on various σ_t surfaces in the southern North Atlantic Ocean, see R. B. Montgomery (1938 a).

4. For a study of the distribution of various chemical constituents of the Atlantic, including an attempt to explain the distribution quantitatively, see G. A. Riley (1951).

5. Charts of mean winds, charts of climatic factors, and similar source material are to be found in *Atlas of Climatic Charts of the Oceans*, prepared by the United States Weather Bureau (1938).

6. Charts showing surface winds, surface currents, and features of navigational interest are available in *Atlas of Pilot Charts, Atlantic Ocean*, prepared by the United States Navy, Hydrographic Office (1950).

7. For a general geography of the Atlantic Ocean, in a popular vein and yet informative and fairly comprehensive, see Gerhard Schott's *Geographie des Atlantischen Ozeans*, published at Hamburg in three editions: 1912, 1926, and 1944. It is the third edition that has been utilized in the present study.

Appendix Three

SIGMA-T VALUES

TABLE A

SHORT TABLE OF SIGMA-T VALUES

Temperature, in degrees Centigrade	Values of σ_t			
	Salinity, in °/oo			
	34	35	36	37
0	27·32	28·13	28·93	—
3	27·11	27·91	28·70	—
6	26·78	27·57	28·36	—
9	26·36	27·14	27·92	—
12	25·83	26·61	27·38	28·16
15	25·22	25·99	26·76	27·53
18	24·53	25·29	26·05	26·82
21	23·75	24·51	25·27	26·04
24	22·91	23·66	24·42	25·18
27	21·99	22·74	23·49	24·25

BIBLIOGRAPHY

AITKEN, JOHN
   1877.   On ocean circulation. Proc. Roy. Soc. Edinburgh, 9:394–400.
ARAGO, F.
   1836.   Ursache der Meeresströmungen. [J. C.] Poggendorffs Ann. d.
           Physik u. Chemie [Leipzig], 37:450–454.
BACHE, A. D.
   1860.   Lecture on the Gulf Stream, prepared at the request of the
           American Association for the Advancement of Science. Am.
           Jour. Sci. and Arts, 2d ser., 30:313–329, plus 3 charts.
BJERKNES, VILHELM
   1898.   Ueber einen hydrodynamischen Fundamentalsatz und seine
           Anwendung besonders auf die Mechanik der Atmosphäre und
           des Weltmeeres. Kongl. Svenska Vetenskaps-Akad. Handl.,
           N.F. [4th ser.], Bd. 31, No. 4. 35 pp. + 5 pp. of figures.
BJERKNES, VILHELM, J. BJERKNES, H. SOLDBERG, and T. BERGERON
   1933.   Physikalische Hydrodynamik. Berlin: J. Springer. Chapter
           XIV: Störung gradliniger Strömungen auf der rotierenden
           Erde. Zyklonenwellen.
BLAGDEN, CHARLES
   1782.   On the heat of the water in the Gulf-stream. Philos. Trans.
           Roy. Soc. London, 71 (Pt. 2, for 1781): 334–344.
BÖHNECKE, GUNTHER
   1936.   Atlas zu: Temperatur, Salzgehalt und Dichte an der Oberfläche
           des Atlantischen Ozeans. In Deutsche Atlantische Expedition
           auf dem Forschungs- und Vermessungsschiff Meteor, 1925–
           1927, Wissenschaftliche Ergebnisse, Bd. 5—Atlas (74 pls.).
           Berlin: Walter de Gruyter & Co.

BOLIN, BERT, and HENRY STOMMEL
1961.   On the abyssal circulation of the world oceans. IV. Origin and rate of circulation of the deep ocean water as determined with the aid of tracers. Deep-Sea Res., 8:95–110.

BRYAN, K.
1963.   A numerical investigation of a nonlinear model of a wind-driven ocean. J. Atm. Sci., 20:594–606.

CARPENTER, W. B.
1874.   Further inquiries on oceanic circulation. Proc. Roy. Geog. Soc., 18:301–407.

CARRIER, G. F., and A. R. ROBINSON
1962.   On the theory of the wind-driven ocean circulation. J. Fluid Mech., 12:49–80.

CHARNEY, JULE G.
1955.   The Gulf Stream as an inertial boundary layer. Proc. Nat. Acad. Sci. Wash., 41:731–740.

CHASE, JOSEPH
1951.   The Bermuda-Azores high pressure cell; its surface wind circulation. (Unpublished manuscript.)

CHURCH, P. E.
1937.   Temperatures of the western North Atlantic from thermograph records. Assoc. d'Océanogr. Phys., Union Géod. et Géophys. Internat., Publ. Sci. No. 4. 32 pp. Liverpool: University of Liverpool.

CONSEIL PERMANENT INTERNATIONAL POUR L'EXPLORATION DE LA MER
1910–1954.   Bull Hydrogr. [Copenhagen], for the years 1908–1949.

DEFANT, ALBERT
1941.   Die absolute Topographie des physikalischen Meeresniveaus und der Druckflächen, sowie die Wasserbewegungen im Atlantischen Ozean. In Deutsche Atlantische Expedition Meteor 1925–1927, Wissenschaftliche Ergebnisse, Bd. 6, 2. Teil, pp. 191–260. Berlin: Walter de Gruyter & Co.

DEFANT, ALBERT, and BJORN HELLAND-HANSEN
1939.   Bericht über die ozeanographischen Untersuchungen im zentralen und östlichen Teil des Nordatlantischen Ozeans im Früh-Sommer 1938 (Internationale Golf-Strom Expedition). Preuss. Akad. d. Wiss., Abhandl., Jahrg. 1939, Phys.-math. Klasse, No. 5. 64 pp.

DUXBURY, A. C.
1963.   An investigation of stable waves along a velocity shear boundary in a two-layer sea with a geostrophic flow regime. Jour. Mar. Res., 21:246–283.

EKMAN, F. WALFRID
1905. On the influence of the earth's rotation on ocean-currents. Arkiv f. Matem., Astr. o. Fysik [Stockholm], Bd. 2, No. 11. 53 pp.
1923. Uber Horizontalzirkulation bei winderzeugten Meeresströmungen. Arkiv f. Matem., Astr. o. Fysik [Stockholm], Bd. 17, No. 26. 74 pp.
1932. Studien zur Dynamik der Meeresströmungen. Gerlands Beitr. z. Geophysik, 36:385–438.
1939. Neuere Ergebnisse und Probleme zur Theorie der Konvektionsströme im Meere. Gerlands Beitr. z. Geophysik, Suppl. 4, Ergebnisse der kosmischen Physik, Bd. 4: Physik der Hydro- und Lithosphäre, pp. 1–74. Leipzig: Akademische Verlagsgesellschaft.

FALLER, A. J.
1960. Further examples of stationary planetary flow patterns in bounded basins. Tellus, 12:160–171.

FELBER, OTTO-HEINR.
1934. Oberflächenströmungen des Nordatlantischen Ozeans zwischen 15° und 50° n. B. Arch. d. Deutsch. Seewarte [Hamburg], Bd. 53, No. 1. 18 pp. + 6 pp. of illustrations.

FERREL, WILLIAM
1882. Popular essays on the movements of the atmosphere. U.S. War Dept., Professional Pap. Signal Serv., No. 12. 59 pp.

FJØRTOFT, RAGNAR
1951. Stability properties of large-scale atmospheric disturbances. In Compendium of meteorology, ed. by Thomas F. Malone. Boston: American Meteorological Society. Pp. 454–463.

FOFONOFF, NICHOLAS P.
1954. Steady flow in a frictionless homogeneous ocean. Jour. Mar. Res., 13:254–262.
1962. The sea, ideas and observations. In Dynamics of ocean currents. New York and London: Pergamon Press. Vol. 1, pp. 323-396.

FORCHHAMMER, GEORG
1865. On the composition of sea-water in the different parts of the ocean. Philos. Trans. Roy. Soc. London, 155:203–262.

FORD, W. L., J. R. LONGARD, and R. E. BANKS
1952. On the nature, occurrence and origin of cold low salinity water along the edge of the Gulf Stream. Jour. Mar. Res., 11:281–293.

FÖRTHMANN, E.
  1934. Uber turbulente Strahlausbreitung. Ingenieur-Archiv, 5:42–
      54.
FRANCIS, J. R. D.
  1951. The aerodynamic drag of a free water surface. Proc. Roy. Soc.
      London, Ser. A, 206:387–406.
FRANKLIN, BENJAMIN
  1786. A letter from Dr. Benjamin Franklin, to Mr. Alphonsus le
      Roy . . . Containing sundry maritime observations. Trans.
      Amer. Philos. Soc., Philadelphia, 2:294–329.
FUGLISTER, F. C.
  1947. Average monthly sea surface temperatures of the western
      North Atlantic Ocean. Pap. Phys. Oceanogr. and Meteor.,
      Vol. 10, No. 2. 25 pp.
  1951a. Annual variations in current speeds in the Gulf Stream Sys-
      tem. Journ. Mar. Res., 10:119–127.
  1951b. Multiple currents in the Gulf Stream System. Tellus, 3:230–
      233.
  1955. Alternative analyses of current surveys. Deep-Sea Res.,
      2:213–229.
  1960. Atlantic Ocean atlas, temperature and salinity profiles and
      data from the International Geophysical Year of 1957–1958.
      Woods Hole Oceanogr. Inst., Atlas Series, 1:1–209.
  1963. Gulf Stream '60. In Progress in Oceanography, Vol. 1, pp.
      265–373.
FUGLISTER, F. C., and L. V. WORTHINGTON
  1951. Some results of a multiple ship survey of the Gulf Stream.
      Tellus, 3:1–14.
GREENSPAN, H. P.
  1962. A criterion for the existence of inertial boundary layers in
      oceanic circulation. Proc. Nat. Acad. Sci., 48:2034–2039.
  1963. A note concerning topography and inertial currents. Jour.
      Mar. Res., 21:147–154.
HAURWITZ, BERNHARD, and H. A. PANOFSKY
  1950. Stability and meandering of the Gulf Stream. Trans. Amer.
      Geophys. Union, 31:723–731.
HELA, ILMO
  1952. The fluctuations of the Florida Current. Bull. Mar. Sci. Gulf
      and Caribbean, 1:241–248.
  1954. The surface current field in the western part of the North
      Atlantic. Bull. Mar. Sci. Gulf and Caribbean, 3:241–
      272.

234 BIBLIOGRAPHY

HELMHOLTZ, HERMANN VON
1868. Uber discontinuirliche Flüssigkeits-Bewegungen. Monatsber. d. Königl. Preuss. Akad. d. Wiss. zu Berlin, 1868, pp. 215–228.

HERRERA Y TORDESILLAS, ANTONIO DE
1601. Historia general de los hechos de los Castellanos en las islas i tierra firme del Mar Oceano. Madrid: Emprenta real, por Iuan Flamenco. 5 vols. in 2.

HIDAKA, KOJI
1949a. Mass transport in ocean currents and lateral mixing. Geophys. Notes, Univ. of Tokyo, Vol. 2, No. 3, pp. 1–4.
1949b. Mass transport in ocean currents and lateral mixing. Jour. Mar. Res., 8:132–136.
1958. Computation of the wind-stresses over the Oceans. Rec. Ocean. Works in Japan, 4:77–123.

HUMBOLDT, ALEXANDER VON
1814. Voyage aux régions équinoxiales. Paris: F. Schoell. 4 vols. folio (1814–1825). 1-65ff. (English ed., Helen Williams [transl.], Personal narrative of travels to the equinoctial regions of the new continent. London: Longman, Hurst, Rees, Orme & Brown [1822]. 1:49–71.).

ICHIYE, TAKASHI
1951. On the variation of oceanic circulation. (1st paper.) Oceanogr. Mag., Tokyo, 3:79–82.

ILYIN, A. N., and V. M. KAMENKOVICH
1963. On the influence of friction on ocean currents. Dokl. Akad. Nauk SSSR, 150:1274–1277.

ISELIN, C. O'D.
1936. A study of the circulation of the western North Atlantic. Pap. Phys. Oceanogr. and Meteor., Vol. 4, No. 4. 101 pp.
1939. The influence of vertical and lateral turbulence on the characteristics of the waters at mid-depths. Trans. Amer. Geophys. Union, 20:414–417.
1940. Preliminary report on long-period variations in the transport of the Gulf Stream System. Pap. Phys. Oceanogr. and Meteor., Vol. 8, No. 1. 40 pp.

ISELIN, C. O'D., and F. C. FUGLISTER
1948. Some recent developments in the study of the Gulf Stream. Jour. Mar. Res., 7:317–329.

JACOBS, W. C.
1942. On the energy exchange between sea and atmosphere. Jour. Mar. Res., 5:37–66.

KELVIN, LORD. See SIR WILLIAM THOMSON

KEULEGAN, G. H.
　1951.　Wind tides in small closed channels. Jour. Res. Nat. Bur. Standards, 46:358–381.

KIRCHER, ATHANASIUS
　1678.　Mundus subterraneus. 3d ed. Amsterdam: Jansson & Sons. 2 vols.

KOHL, J. G.
　1868.　Geschichte des Golfstroms und seiner Erforschung. Bremen: C. E. Müller.

KUO, HSIAO-LAN
　1949.　Dynamic instability of two-dimensional nondivergent flow in a barotropic atmosphere. Jour. Meteor., 6:105–122.

LAMB, SIR HORACE
　1932.　Hydrodynamics. 6th ed. Cambridge, England: Cambridge University Press.

LAPLACE, P. S.
　1778.　Recherches sur plusieurs points du système du monde. Mém. de l'Acad. Roy. d. Sci. de Paris, année 1775. (Also in his (Œuvres complètes, 9 1893:88, 187; and, in revised form, in Mécanique céleste, Livre 4me [1799], Chapter I.)

LAVAL, [ANTOINE F.]
　1728.　Voyage de la Louisiane. Paris: J. Mariette.

LESCARBOT, MARC
　1612.　Histoire de la Nouvelle-France. 2d ed. Paris: Jean Milot. 2 vols.

LINEYKIN, P. C.
　1955.　On the determination of the thickness of the baroclinic layer of fluid heated uniformly above and non-uniformly from below. Dokl. Ak. Nauk SSSR, 101:461.

LIPPS, F. B.
　1963.　Stability of jets in a divergent barotropic fluid. Jour. Atm. Sci., 20:120–129.

LONGUET-HIGGINS, M. S., M. E. STERN, and HENRY STOMMEL
　1954.　The electric field induced by ocean currents and waves, with applications to the method of towed electrodes. Pap. Phys. Oceanogr. and Meteor., Vol. 13, No. 1. 37 pp.

MALKUS, W. V. R.
　1953.　A recording bathypitotmeter. Jour. Mar. Res., 12:51–59.

MAURY, M. F.
　1859.　The physical geography of the sea. Rev. ed. New York: Harper & Bros.

MIYAZAKI, M.
　1952.　Notes on the theory of the wind-driven oceanic circulation.

Oceanogr. Mag., Tokyo, 4:31–36.

MOHN, HENRIK
1885. Die Strömungen des europäischen Nordmeeres. Petermanns Mitt., Ergänzungshefte No. 79. 20 pp. + 4 pp. of maps. Gotha: Justus Perthes.

MONTGOMERY, R. B.
1936a. Computation of the transport of surface-water due to the wind-system over the North Atlantic. Trans. Amer. Geophys. Union, 17:225–229.

1936b. On the momentum transfer at the sea surface. III. Transport of surface water due to the wind system over the North Atlantic. Pap. Phys. Oceanogr. and Meteor., Vol. 4, No. 3, pp. 23–30.

1938a. Circulation in upper layers of southern North Atlantic deduced with use of isentropic analysis. Pap. Phys. Oceanogr. and Meteor, Vol. 6, No. 2. 55 pp.

1938b. Fluctuations in monthly sea level on eastern U.S. coast as related to dynamics of western North Atlantic Ocean. Jour. Mar. Res., 1:165–185.

1939. Ein Versuch, den vertikalen und seitlichen Austausch in der Tiefe der Sprungschicht im äquatorialen Atlantischen Ozean zu bestimmen. Ann. d. Hydrogr. u Marit. Meteor., 67:242–246.

1940. The present evidence on the importance of lateral mixing processes in the ocean. Bull. Amer. Meteor. Soc., 21:87–94.

1941a. Sea level difference between Key West and Miami, Florida. Jour. Mar. Res., 4:32–37.

1941b. Transport of the Florida Current off Habana. Jour. Mar. Res., 4:198–220.

1947. Submarine tubes for levelling. Nature [London], 159:408.

MONTGOMERY, R. B., and ERIK PALMEN
1940. Contribution to the question of the Equatorial Counter Current. Jour. Mar. Res., 3:112–133.

MONTGOMERY, R. B., and M. J. POLLAK
1942. Sigma-T surfaces in the Atlantic Ocean. Jour. Mar. Res., 5:20–27.

MOORE, D. W.
1963. Rossby waves in ocean circulation. Deep-Sea Res., 10:735–748.

MORGAN, G. W.
1956. On the wind-driven ocean circulation. Tellus, 8:301–320.

MUNK, W. H.
1947. A critical wind speed for air-sea boundary processes. Jour. Mar. Res., 6:203–218.

1950. On the wind-driven ocean circulation. Jour. Meteor., 7:79–93.

MUNK, W. H., and G. F. CARRIER
1950. The wind-driven circulation in ocean basins of various shapes. Tellus, 2:158–167.

MUNK, W. H., G. W. GROVES, and G. F. CARRIER
1950. Note on the dynamics of the Gulf Stream. Jour. Mar. Res., 9:218–238.

MURRAY, K. M.
1952. Short period fluctuations of the Florida Current from geomagnetic electrokinetograph observations. Bull. Mar. Sci. Gulf and Caribbean, 2:360–375.

NEUMANN, GERHARD
1940. Die ozeanographischen Verhältnisse an der Meeresoberfläche im Golfstromsektor nördlich und nordwestlich der Azoren. Ann. d. Hydrogr. u. Marit. Meteor., Bd. 68, Beiheft z. Juniheft, Lieferung 1. 87 pp.

1948. Uber den Tangentialdruck des Windes und die Rauhigkeit der Meeresoberfläche. Zeitschr. f. Meteor., 2:193–203.

1952. Some problems concerning the dynamics of the Gulf Stream. Trans. N. Y. Acad. Sci., 2d ser., 14:283–291.

PARR, A. E.
1937a. A contribution to the hydrography of the Caribbean and Cayman Seas, based upon the observations made by the research ship Atlantis, 1933–34. Bull. Bingham Oceanogr. Collection [Yale Univ.], Vol. 5, Art. 4. 110 pp.

1937b. Report on hydrographic observations at a series of anchor stations across the Straits of Florida. Bull. Bingham Oceanogr. Collection [Yale Univ.], Vol. 6, Art. 3. 62 pp.

PETER MARTYR OF ANGHIERA
1577. The decades of the ocean. English translation in Richarde Eden (ed. and transl.), The history of trauayle in the West and East Indies . . . London: Richarde Iugge. Dec. III. Book 6, p. 127.

PETERS, H., and J. BICKNELL
1936. (Unpublished data on file at the Massachusetts Institute of Technology.)

PHILLIPS, N. A.
1951. A simple three-dimensional model for the study of large-scale extratropical flow patterns. Jour. Meteor., 8:381–394.

PILLSBURY, J. E.
1891. The Gulf Stream . . . Rept. Supt., U.S. Coast and Geod. Surv., for year ending June, 1890, pp. 459–620 (= Appen. 10).

POWNALL, THOMAS
  1787.   Hydraulic and nautical observations on the currents in the
          Atlantic Ocean . . . London: Printed for Robert Sayer. 17 pp.
PROUDMAN, JOSEPH
  1953.   Dynamical oceanography. London: Methuen & Co., Ltd.; and
          New York: Wiley & Sons, Inc.
QUENEY, PAUL
  1950.   Adiabatic perturbation equations for a zonal atmospheric cur-
          rent. Tellus, 2:35–51.
REID, R. O.
  1948a. The equatorial currents of the eastern Pacific as maintained
          by the stress of the wind. Jour. Mar. Res., 7:74–99.
  1948b. A model of the vertical structure of mass in equatorial wind-
          driven currents of a baroclinic ocean. Jour. Mar. Res., 7:304–
          312.
RENNELL, JAMES
  1832.   An investigation of the currents of the Atlantic Ocean (+Atlas
          of 5 [10] charts). London: Publ. for Lady Rodd by J. G. &
          F. Rivington.
REYNOLDS, OSBORNE
  1883.   An experimental investigation of the circumstances which
          determine whether the motion of water shall be direct or
          sinuous, and of the law of resistance in parallel channels.
          Philos. Trans. Roy. Soc. London, 174:935–982.
RICHARDS, F. A., and A. C. REDFIELD
  1955.   Oxygen-density relationships in the western North Atlantic.
          Deep-Sea Res., 2:182–199.
RILEY, G. A.
  1951.   Oxygen, phosphate, and nitrate in the Atlantic Ocean. Bull.
          Bingham Oceanogr. Collection [Yale Univ.], Vol. 13, Art. 1.
          128 pp.
ROBINSON, A. R. ed.
  1963.   On the wind-driven ocean circulation. New York: Blaisdell
          Press, 161 pp.
ROBINSON, A. R., and HENRY STOMMEL
  1959.   The oceanic thermocline and the associated thermohaline cir-
          culation. Tellus, 11:295–308.
ROBINSON, A. R., and PIERRE WELANDER
  1963.   Thermal circulation on a rotating sphere; with application to
          the oceanic thermocline. J. Mar. Res., 21:25–38.
ROSSBY, CARL-GUSTAF
  1936a. Dynamics of steady ocean currents in the light of experimental

fluid mechanics. Pap. Phys. Oceanogr. and Meteor., Vol. 5, No. 1. 43 pp.

1936b. On the momentum transfer at the sea surface. I. On the frictional force between air and water and on the occurrence of a laminar boundary layer next to the surface of the sea. Pap. Phys. Oceanogr. and Meteor., Vol. 4, No. 3, pp. 1–20.

1947. On the distribution of angular velocity in gaseous envelopes under the influence of large-scale horizontal mixing processes. Bull. Amer. Meteor. Soc., 28:53–68.

1951. On the vertical and horizontal concentration of momentum in air and ocean currents. Tellus, 3:15–27.

ROSSBY, CARL-GUSTAF, and R. B. MONTGOMERY
1935. The layer of frictional influence in wind and ocean currents. Pap. Phys. Oceanogr. and Meteor., Vol. 3, No. 3. 101 pp.

[ROSSBY, CARL-GUSTAF, and] STAFF OF METEOROLOGY DEPARTMENT, UNIVERSITY OF CHICAGO
1947. On the general circulation of the atmosphere in middle latitudes. Bull. Amer. Meteor. Soc., 28:255–280.

SABINE, SIR EDWARD
1825. An account of experiments to determine the figure of the earth. London: J. Murray.

SANDSTROM, J. W., and BJORN HELLAND-HANSEN
1903. Uber die Berechnung von Meeresströmungen. Repts. Norweg. Fish. and Mar. Investigations, Vol. 2, No. 4. 43 pp.

SARKISYAN, A. C.
1954. The calculation of stationary wind currents in an ocean. Izvestia, Akad. Nauk SSR, Ser. Geofiz., 6:554–561.

SCHOTT, GERHARD
1944. Geographie des Atlantischen Ozeans. 3d ed. Hamburg: C. Boysen.

SCHROEDER, E., H. M. STOMMEL, D. W. MENZEL, and W. H. SUTCLIFFE, JR.
1959. Climatic stability of eighteen degree water at Bermuda. J. Geophys. Res., 64:363–366.

SHEPPARD, P. A., HENRY CHARNOCK, and J. R. D. FRANCIS
1952. Observations of the westerlies over the sea. Quart. Jour. Roy. Meteor. Soc. [London], 78:563–582.

SHEPPARD, P. A., and M. H. OMAR
1952. The wind stress over the ocean from observations in the trades. Quart. Jour. Roy. Meteor. Soc. [London], 78:583–589.

SOULE, F. M.
1950. Physical oceanography of the Grand Banks region and the

240　　　　　　　　　　　　　　　　BIBLIOGRAPHY

  Labrador Sea in 1950. U.S. Treas. Dept. Coast Guard Bull. No. 36, pp. 61–93.

SOULE, F. M., P. A. MORRILL, and A. P. FRANCESCHETTI
 1961. Physical oceanography of the Grand Banks region and the Labrador Sea in 1960. U.S. Coast Guard Bull., 46:31–114.

STARR, V. P.
 1953. Note concerning the nature of large-scale eddies in the atmosphere. Tellus, 5:494–498.

STERN, M.
 1961. The stability of thermoclinic jets. Tellus, 13:503–508.

STOCKMANN, W. B.
 1946. Equations for a field of total flow induced by the wind in a nonhomogeneous sea. Comptes Rendus (Doklady) de l'Acad. d. Sci. de l'URSS, n.s., 54:403–406.

STOMMEL, H. M.
 1948. The westward intensification of wind-driven ocean currents. Trans. Amer. Geophys. Union, 29:202–206.

 1951. Determination of the lateral eddy diffusivity in the climatological mean Gulf Stream. Tellus, 3:43.

 1953. Examples of the possible role of inertia and stratification in the dynamics of the Gulf Stream System. Jour. Mar. Res., 12:184–195.

 1955a. Discussion at the Woods Hole Convocation, June, 1954. Jour. Mar. Res., 14:504–510.

 1955b. Lateral eddy viscosity in the Gulf Stream System. Deep-Sea Res., 3:88–90.

 1963. Varieties of oceanographic experience. Science, 139:572–576.

 1964. Summary charts of the mean dynamic topography and current field at the surface of the ocean, and related functions of the mean wind-stress. Oceanographic Papers (Hidaka Jubilee Volume), in press.

STOMMEL, H. M., and A. B. ARONS
 1960a. On the abyssal circulation of the world ocean. I. Stationary planetary flow patterns on a sphere. Deep-Sea Res., 6:140–154.

 1960b. II. An idealised model of the circulation pattern and amplitude in oceanic basins. Deep-Sea Res., 6:217–233.

STOMMEL, H. M., A. B. ARONS, and A. J. FALLER
 1958. Some examples of stationary planetary flow patterns in bounded basins. Tellus, 10:179–187.

STOMMEL, H. M., and G. VERONIS
 1957. Steady convective motion in a horizontal layer of fluid heated uniformly above and non-uniformly from below. Tellus,

9:401–407.

STOMMEL, H. M., W. S. VON ARX, DONALD PARSON, and W. S.
RICHARDSON
1953.   Rapid aerial survey of Gulf Stream with camera and radiation
thermometer. Science, 117:639–640.

STOMMEL, H. M., and J. WEBSTER
1962.   Some properties of the thermocline equations in a subtropical
gyre. J. Mar. Res., 20:42–56.

STRICKLAND, CAPT. WILLIAM
1802.   On the use of the thermometer in navigation. Trans. Amer.
Philos. Soc., Philadelphia, 5:90–103.

SVERDRUP, H. U.
1947.   Wind-driven currents in a baroclinic ocean; with application
to the equatorial currents of the eastern Pacific. Proc. Nat.
Acad. Sci. Wash., 33:318–326.

SVERDRUP, H. U., M. W. JOHNSON, and R. H. FLEMING
1942.   The oceans: their physics, chemistry and general biology. New
York: Prentice-Hall.

SWALLOW, J. C., and L. V. WORTHINGTON
1961.   An observation of a deep countercurrent in the Western North
Atlantic. Deep-Sea Res., 8:1–19.

THEVET, ANDRE
1575.   La Cosmographie universelle. Paris: P. L'Huilier. 2 vols.
(Pagination continuous).

THOMSON, SIR [CHARLES] WYVILLE
1874.   The depths of the sea. 2d ed. London: Macmillan and Co.
Chapter VIII: The Gulf-Stream.

THOMSON, SIR WILLIAM [later, LORD KELVIN]
1871.   Hydrokinetic solutions and observations. Philos. Mag. and
Jour. Sci., 4th ser., Vol. 42, Pt. 2, pp. 362–377.

TOLLMIEN, WALTER
1926.   Berechnung turbulenter Ausbreitungsvorgänge. Zeitschr. f.
Angew. Math. u. Mech., 6:468–478.

TOLSTOY, IVAN
1951.   Submarine topography in the North Atlantic. Bull Geol. Soc.
Amer., 62:441–450.

UDA, M.
1938.   Hydrographical fluctuation in the northeastern sea-region
adjacent to Japan of North Pacific Ocean. Jour. Imper.
Fisheries Sta. Japan, 9:64–85.

UNITED STATES NAVY, HYDROGRAPHIC OFFICE
1946.   Current atlas of the North Atlantic Ocean. Hydrogr. Office

Misc. No. 10,688.
1950.   Atlas of pilot charts, Atlantic Ocean. Hydrogr. Office Publ. No. 576.

UNITED STATES WEATHER BUREAU
1938.   Atlas of climatic charts of the oceans. Washington, D.C.
[1942–1955]. Historical weather maps, Northern Hemisphere, sea level. For period from January, 1899, through June, 1939. Continued as: Synoptic weather maps, daily series, Northern Hemisphere, sea level and 500 mb. Charts with synoptic data tabulations. For period from January, 1949, through June, 1953. Washington, D.C.: U.S. Weather Bureau.
1952.   Normal weather maps. U.S. Weather Bureau, Tech. Publ. No. 21.

VAN DORN, W. C.
1953.   Wind stress on an artificial pond. Jour. Mar. Res., 12:249–276.

VARENIUS, BERNARDUS [BERNHARD VAREN]
1681.   Geographia generalis. 2d ed. Cambridge, England: Henry Dickinson.

VAUGHAN, THOMAS W., and OTHERS
1937.   International Aspects of Oceanography. Washington, D.C.: National Academy of Sciences.

VERONIS, GEORGE
1963a.   On inertially-controlled flow patterns in a beta-plane ocean. Tellus, 15:59–66.
1963b.   An analysis of wind-driven ocean circulation with a limited number of Fourier components. J. Atm. Sci., 20:577–593.

VERONIS, GEORGE, and G. W. MORGAN
1955.   A study of the time-dependent wind-driven ocean circulation in a homogeneous, rectangular ocean. Tellus, 7:232–242.

VERONIS, GEORGE, and HENRY STOMMEL
1956.   The action of variable wind stresses on a stratified ocean. Jour. Mar. Res., 15:43–75.

VOLKMANN, G.
1962.   Deep current measurements in the western North Atlantic. Deep-Sea Res., 9:493–500.

VON ARX, W. S.
1950.   An electromagnetic method for measuring the velocities of ocean currents from a ship under way. Pap. Phys. Oceanogr. and Meteor., Vol. II, No. 3. 62 pp.

VON ARX, W. S., D. BUMPUS, and W. S. RICHARDSON
1955.   On the fine structure of the Gulf Stream front. Deep-Sea Res., 3:46–65.

VOSSIUS, ISAAC
1663. De motu marium et ventorum. The Hague: Adrian Vlacq.
WARREN, B. A.
1963. Topographic influences on the path of the Gulf Stream. Tellus, 15:167–183.
WEBSTER, F.
1961a. The effect of meanders on the kinetic energy balance of the Gulf Stream. Tellus, 13:392–401.
1961b. A description of Gulf Stream meanders off Onslow Bay. Deep-Sea Res., 8:130–143.
WELANDER, PIERRE
1959. An advective model of the ocean thermocline. Tellus, 11:309–318.
WERTHEIM, GUNTHER K.
1954. Studies of the electric potential between Key West, Florida, and Havana, Cuba. Trans. Amer. Geophys. Union, 35:872–882.
WORTHINGTON, L. V.
1954a. A preliminary note on the time scale in North Atlantic circulation. Deep-Sea Res., 1:244–251.
1954b. Three detailed cross-sections of the Gulf Stream. Tellus, 6:116–123.
1959. The 18° water in the Sargasso Sea. Deep-Sea Res., 5:297–305.
WÜST, GEORG
1924. Florida- und Antillenstrom, eine hydrodynamische untersuchung. Veröff. d. Inst. f. Meereskunde an d. Univ. Berlin, N.F., Reihe A: Geogr.-naturwiss., Heft 12. 48 pp + 1 p. of illustrations.
1936. Kuroshio und Golfstrom. Veröff. d. Inst. f. Meereskunde an d. Univ. Berlin, N.F., Reihe A: Geogr.-naturwiss., Heft 29. 70 pp. + map.
WÜST, GEORG, and ALBERT DEFANT
1936. Atlas zur Schichtung und Zirkulation des Atlantischen Ozeans. Schnitte und Karten von Temperatur, Salzegehalt, und Dichte. In Deutsche Atlantische Expedition auf dem Forschungs- und Vermessungsschiff Meteor, 1925–1927, Wissenschaftliche Ergebnisse, Bd. 6—Atlas (103 pls.). Berlin: Walter de Gruyter & Co.
ZÖPPRITZ, K.
1878. Hydrodynamische Probleme in Beziehung zur Theorie der Meeresströmungen. Ann. d. Physik u. Chemie, N.F., 3:582–607.

INDEX